Solid State Devices, 1973

Solid State Devices, 1973

The six invited papers presented at the Third European Solid State Device Research Conference (ESSDERC) sponsored by the European Physical Society, The Institute of Physics, the Deutsche Physikalische Gesellschaft, Region 8 of the Institute of Electrical and Electronics Engineers and the Nachrichtentechnische Gesellschaft im VDE, held in Munich, 18–21 September 1973

Conference Series Number 19

The Institute of Physics
London and Bristol

ISBN O 85498 109 8

The Third European Solid State Device Research
Conference was sponsored by the European Physical
Society, The Institute of Physics, the Deutsche
Physikalische Gesellschaft, Region 8 of the Institute of
Electrical and Electronics Engineers and the
Nachrichtentechnische Gesellschaft im VDE.

Organizing Committee:
    Chairman: H Weiss
    Vice-Chairman: R Muller
    Secretary: E Preuss

Program Committee: H Beneking, J Borel, O G Folberth,
    J Hesse, W Heywang, C Hilsum, F Paschke,
    L J Tummers, T Wessel-Berg.
Corresponding members: G H Schwuttke, F M Smits,
    H K Zaininger.

Honorary Editor:
    H Weiss

Published by The Institute of Physics, 47 Belgrave
Square, London SW1X 8QX, England, and 23 Marsh Street,
Bristol BS1 4BT, England, in association with the American
Institute of Physics, 335 East 45th Street, New York, NY
10017, USA.

Set in 11/12 pt Monotype Imprint, and printed in Great
Britain by Adlard and Son, Ltd, Dorking, Surrey.

Contents

v

Preface

The Third European Solid State Device Research Conference (ESSDERC 1973) was held in Munich from 18 to 21 September 1973 and was a continuation of the two preceding conferences held in Munich in 1971 and in Lancaster in 1972. Since the first European conference of this type in 1971 there has been an increasing interest in the subject, as is demonstrated by the large number of contributed papers and participants. In fact 530 people from 18 countries attended in 1973, coming from Europe, America and Asia.

The six invited and eighty contributed papers gave a survey of the progress in the whole area of solid state devices, from materials technology to novel devices and problems of integration and miniaturization. A few main topics may be mentioned: faster and smaller MOS transistors; luminescence and laser diodes for visible light with better stability and reliability; thin mono-crystalline layers instead of bulk material for devices with higher speed and less power consumption; and memory devices.

Looking back over the conference we can say that although no basically new devices were announced, much progress had been made in the fields of experimental methods and the theoretical understanding of the devices necessary to reach the goals: 'faster', 'a higher degree of integration' and 'less power'.

The present volume contains the six invited papers and covers various areas of device research and application. I would like to thank the invited speakers for the ready delivery of their manuscripts.

H Weiss
October 1973

Semiconductor optoelectronics

M H Pilkuhn

Physikalisches Institut, Universität Stuttgart,
7000 Stuttgart 1, Wiederholdstr 13, West Germany

Abstract

Semiconductor light emitters are discussed in this paper in connection with new display technologies and optical communication. After a general review of semiconductor display materials, the III–V compound ternary alloys are treated in detail. Ga(AsP) which has the most advanced technology today receives large attention. Improvements for indirect transitions in Ga(AsP) have been recently obtained by nitrogen doping.

After a review of the present status of GaP LEDs, the role of isoelectronic traps and of Auger recombination is pointed out. Problems of bulk crystal growth are discussed.

In optical communication, the glass fibre and semiconductor laser technologies play an important role and have experienced dramatic improvements. The attenuation in glass waveguides seems to be mainly limited by light scattering after elimination of absorptive losses. For semiconductor lasers, continuous operation at room temperature has been achieved with double heterostructures. The lowest threshold current densities are obtained by a separate electrical and optical confinement. Recent progress in stripe geometry structures and in the degradation problem appears to be quite significant. The influence of strain on the slow degradation of GaAs lasers is pointed out. Finally, semiconductor lasers in integrated optics are briefly discussed.

1. Introduction

A concise and clear definition of the field of 'optoelectronics' is quite difficult to attain. Conversion of electrical into optical signals, and vice versa, is certainly only one aspect of the field, but not a general description. All electrical devices capable of generating, modulating and detecting light, as well as elements of light transmission, optical memories etc, have to be included in optoelectronics. A list of some modern topics and elements which appear to be of special interest is given in table 1.

Table 1. Selected elements and topics of modern solid-state opto-electronics

Emitters and displays	Visible displays: alphanumeric, digital computer displays. Large area, flat panel displays. Light-emitting diodes, indicator lights. Infrared emitters.
Lasers	Semiconductor lasers and solid-state (Nd) lasers for optical communication. Tunable infrared lasers for pollution measurements. Lasers as intense and coherent radiation sources for general purposes.
Detectors	Photodiodes, avalanche detectors, photoconductors. Various detector arrays. Special infrared detectors. Solar cells, sensors, semiconductor cathodes. Infrared and low level imaging. Image intensifiers. Photo-transistors, optical couplers.
Electro-optic modulators	Nonlinear optical devices.
Integrated optics	Waveguides, optical fibres.
Optical information storage	Holographic memories.

Certainly, the field as such is not new and comprises some traditional areas like incandescent lamps, cathode ray tubes (CRT), gas discharge displays or photomultipliers, which have not lost their attractiveness for a large number of modern applications.

Two of the very important new areas which have developed at an increasingly rapid rate and which will receive particular attention in this review paper are (i) *new display technologies* and (ii) *optical communication*. The major emphasis in this paper will lie on the semiconductor aspects of light emission, whereas detectors and modulators will not be discussed.†

† A recent collection of review papers on 'Semiconductor Light Emitters and Detectors' may be found in the special issue of *J. Luminescence* **7**, 1973 (Proceedings of the International Symposium held at Pugnochiuso, Italy, Sept 1972).

New semiconductor materials and technologies are needed for these areas of optoelectronics, in particular for light generation and modulation. It seems important to point out that they will not find the established silicon technology to be a *competitor* but rather a *partner*, namely on the electronics side of *opto-electronics*. This leads to an essential conclusion: in order to faciliate this partnership, optoelectronic devices should be compatible with silicon integrated circuits (SIC). For anticipating which optoelectronic technology (of the many developed now) will eventually survive, SIC compatibility will be a decisive criterion.

Although optoelectronic components are produced in large numbers today, many of the technologies have to be considered still as being underdeveloped. Significant progress through the converging efforts of solid-state physicists, materials scientists and electrical engineers may be anticipated for the years to come.

2. New display technologies

The largest number of displays today exist in the form of CRT-TV displays giving an estimated total display area of 50 km² in the world (Loebner 1973). Although the number of CRT displays will certainly increase considerably, the largest demand is anticipated for small and intermediate displays, for which a seemingly unlimited number of applications as alphanumeric, digital or graphic displays appears. For the end of the century, a worldwide total display area of 2000 km² is anticipated (Loebner 1973), a demand which will implement probably several new technologies.

2.1. Comparison of technologies

Displays can be classified as (i) light-emitting (active) and (ii) light-modulating (passive) displays. The first group comprises light-emitting diodes (LED), tungsten lamps, CRT, EL DC or AC phosphors and gas discharge (plasma panel) displays. Liquid crystal displays (LCD), electrophoretic, electrochromic and ferroelectric displays are examples of the second group. More information may be found in the paper of van Houten presented at this conference (van Houten 1974), and in a special issue of the Proceedings of the IEEE (van Houten 1973) on 'new materials for display devices'. For small and inter-mediate size displays, the LED, LCD, gas discharge, and DC EL phosphor panels can be viewed as major competitors at the present time, if the traditional tungsten lamps are left out. A schematic diagram showing the relative potential of the four technologies as a function of the number of characters, N, of the display is shown in figure 1. For a small number of characters, obviously LEDs have the largest potential (including $N=1$, which is the indicator light). LCDs, gas discharge, and phosphor panels show a better potential for a large number of characters. There are, of course, many ways for comparing the different

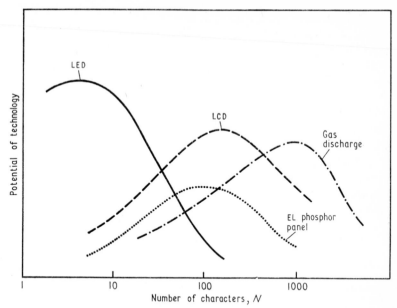

Figure 1.   Relative potential of display technologies as a function of the number of characters (Triebwasser 1973).

Table 2.   Comparison of technologies

Technology	LED	LCD	Gas discharge panel	DC EL panel
Appearance	bright	depending on illumination	bright	bright
SIC compatibility	yes	yes	no	no
Reliability, life	very good	questionable	good	satisfactory
Power consumption	high	very low	high	low–intermediate
Cost per element	high	low	low	low
Switching speed	high	very low	sufficient	sufficient

technologies, and some of them are listed in table 2. Other criteria are: operating voltage, threshold for switching, inherent memory, element size and range of colours. Cost, as usual, will be one of the most important criteria. For this purpose, however, the total cost of the display including addressing electronics and not just the cost per element should be considered. For LEDs, for instance, the cost per element is high, but the total display cost may be low because of the SIC compatibility. For gas discharge panels it is just the other way around, because they are not SIC compatible.

2.2. Semiconductor display technologies

A large number of potential semiconductors exist (compare Dean 1973a) which, in principle, should make it possible to realize almost every colour with LEDs.

However, for bandgaps over 2·2 eV it becomes difficult or impossible to produce both n- and p-type conductivity which is obviously essential for the fabrication of diodes. Other devices like Schottky barriers, heterojunctions and avalanche and tunnel devices have been explored to circumvent this problem and they try to use material of one conductivity type (or insulating material) only. All of them, however, are still at a laboratory stage leaving the pn junction diode as the only technologically developed and commercially attractive device in the

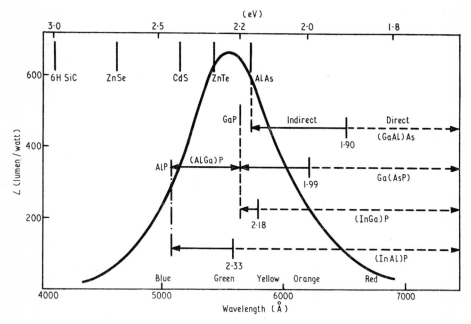

Figure 2. Luminous efficiency (*L*) of the human eye. The energy gaps of some semiconductor display materials are included, in particular the energy ranges covered by various III–V ternary alloy compounds.

field. This reduces the choice of semiconductors down to very few, and figure 2 shows the energy gaps for some materials of interest for visible LEDs in connection with the sensitivity curve of the human eye. The energy gap is the upper limit of the energy of the transition which can be accommodated in the material. Of course, semiconductors with a large gap (eg GaN, ZnS) can also have radiative transitions in the visible range.

   There are many ways of making red LEDs using various ternary alloy systems of III–V compounds and also using GaP (doped with O and Zn). Reasonable technologies have been developed for GaP, Ga(AsP), and (GaAl)As. The III–V compounds, in fact, have no serious competitors right now for producing

LEDs in the colour range from red to green. The aluminium compounds AlP and AlAs will presumably retain importance in connection with ternary alloy systems only, since the compounds themselves suffer from instability. The II–VI compounds ZnSe, CdS and ZnTe are in principle ideal candidates for the colour range green to blue; however, luminescent pn junctions can not in practice be made from them. For the green-emitting GaP (and related ternary compounds) they are no competition.

Today, no really good technology for blue-emitting diodes is in sight. The most promising one existing is GaN, where transitions to deep levels can generate blue light (see Pankove 1973 for a review of recent work on GaN). SiC can be made both n- and p-type, and the energy gap of the 6H-polytype, which is one of the more common ones, is included in figure 2. There are, however, many difficulties in the high temperature technology of SiC diodes which severely limit their large scale application. Blue LEDs will be in the same difficult situation as red ones: since the sensitivity of the eye drops in these regions, they have to be much more efficient in order to compete with green-emitting diodes.

Alternative materials for the complete colour range, blue to red, exist as II–IV–V$_2$ compounds (eg, CdSiP$_2$, ZnSiAs$_2$, ZnSiP$_2$, ZnGeP$_2$, CdGeP$_2$) and I–III–VI$_2$ compounds (eg, CuAlSe$_2$, AgGaS$_2$, AgInS$_2$, AgGaSe$_2$). These compounds (whose energy gaps are not included in figure 2) have been investigated only very recently (Shay 1972) and do seem to show interesting luminescence properties in a few cases. One should remember, though, that they will always have to compete with already existing GaP or Ga(AsP) technologies and this leaves them only a small chance of success. In summary, ternary III–V alloys and GaP are the only optoelectronic semiconductors of serious interest for red to green light-emitting diodes, and their potential and problems will be reviewed in more detail.

2.3. III–V compound alloy systems

The extensive research carried out on the ternary alloy systems of the III–V compounds GaAs, AlAs, GaP and InP was motivated by the hope that all these systems should exhibit quantum efficiencies similar to that of GaAs, as long as they had a 'direct' energy gap. Since the external quantum efficiency of the best GaAs diodes (Si-doped and dome shaped) could be made as high as 28% (Ashley and Strack 1969) at room temperature, expectations for systems like (GaAl)As, Ga(AsP), (InGa)P and (InAl)P were naturally very high. All of these systems have a transition from a direct gap to indirect gap (compare figure 2) at which the quantum efficiency drops very rapidly. This was first found experimentally for Ga(AsP) (Pilkuhn and Rupprecht 1965, Maruska and Pankove 1967) and later for (GaAl)As (Kressel *et al* 1969) and (InGa)P (Onton *et al* 1971). If n_Γ denotes the concentration of electrons in the central Γ-minimum ($k=0$), and n_X that in the subsidiary X-minimum, from which indirect transitions with negligibly small quantum efficiency are assumed, the total efficiency drop can be described by the relation

$$\eta_{\text{ext}} = f \left[1 + \frac{\tau_r}{\tau_{nr}} \left(1 + \frac{n_X}{n_\Gamma} \right) \right]^{-1}.$$

Here, τ_r = radiative lifetime, τ_{nr} = nonradiative lifetime for the direct transition and f = geometry factor. The efficiency drop takes place at energies where the eye sensitivity is still increasing, and the result of these two effects can be best seen in terms of brightness B (in ft L) divided by current density j (in A cm^{-2}), a ratio which can be used as a 'figure of merit' for LEDs:

$$\frac{B}{j} = 1150 \, \frac{\eta_{\text{ext}} L}{\lambda} \, (\text{ft L A}^{-1} \, \text{cm}^2),$$

where L = luminous efficiency of the eye in lumens/watt, λ = wavelength in μm.

A theoretical plot of the figure of merit, B/j, as a function of photon energy is shown in figure 3, assuming $\tau_r = \tau_{nr}$ for the direct gap recombination, in other words that the 'internal' quantum efficiency

$$\eta_i = \left(1 + \frac{\tau_r}{\tau_{nr}} \right)^{-1} = 0 \cdot 5 \simeq 50\%.$$

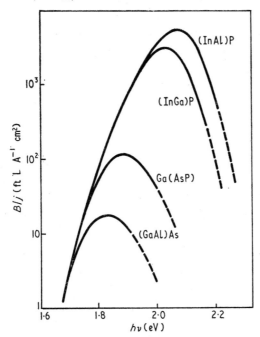

Figure 3. Theoretical dependence of brightness (divided by current density) on photon energy for various ternary alloy compounds (Archer 1972).

Since (InAl)P has the direct–indirect crossover point of the highest energy (at about 2·33 eV), it should be the best of the alloy systems of figure 3.   Unfortunately, its technology is the most unpleasant one and not yet mastered even in the laboratory.   Much laboratory work is going on in the case of (InGa)P, but this system too has been plagued by a difficult materials technology.   It has been produced for pn junction luminescence by melt growth (Onton and Lorenz 1971), by vapour phase epitaxy (VPE) (Sigai *et al* 1973), and by liquid phase epitaxy (LPE) (Hakki 1971).   The problems that arise in the material synthesis do not seem to be quite as severe as for (InAl)P.   In both cases they originate from the relatively large lattice mismatch between the end components and from the disparity between the free energies of formation of the end components.   The problem of finding a suitable substrate for epitaxy can be solved in an elegant way for $In_{0.5}Ga_{0.5}P$ (direct gap of 1·9 eV) by using GaAs which has the same lattice constant as that particular alloy (Kressel *et al* 1973).

The performance of pn junction luminescence of VPE (InGa)P (Nuese *et al* 1973) has recently become comparable to the best data reported for LPE material (Archer 1972).   At an emission wavelength of 6170 Å, quantum efficiencies of 6×10^{-4} have been reached at room temperature, leading to a best brightness value (per current density) of about 320 $(ft\,L\,A^{-1}\,cm^2)$.   Encapsulation and proper shaping of diodes can improve these values easily by a factor of 2.

The efficiencies of (InGa)P diodes seem to be limited by infrared luminescence and by radiationless contributions to the recombination, probably influenced by dislocations, Zn_3P_2 precipitates and other defects.   In spite of the difficulties in materials technology, the system will probably remain attractive, because direct energy gaps up to 2·2 eV can be reached.   Progress through improved materials and diode technology, especially through nitrogen doping (isoelectronic traps, higher efficiency on the indirect gap side) can be expected.

Let us now turn to the lowest curve in figure 3, the alloy system (GaAl)As. The good lattice constant matching between GaAs and AlAs makes the material synthesis very simple: LPE layers of (GaAl)As can be easily grown on GaAs substrates (Woodall *et al* 1969).   This has become the standard technology for the fabrication of heterostructure lasers.   However, for red LEDs the system is not very favourable, as may be recognized from figure 2 (lowest theoretical brightness values).   External quantum efficiencies for direct energy gaps are quite high and have reached room temperature values of 6–10% (near 7500 Å) by applying composition gradients, by epoxy-coating, or by using special heterostructures (GaAs : Si–(GaAl)As : Zn) in order to reduce, for instance, reabsorption losses.   Because of the relatively long wavelengths, the brightness values are not too good.   Economical VPE synthesis of (GaAl)As has not been achieved.

The most advanced ternary alloy LED technology is that of Ga(AsP), which according to figure 3 shows an intermediate potential for brightness.   The standard fabrication process is VPE on GaAs substrates, whereby the change in lattice constant is matched by a gradual change in composition (Craford and Groves 1973). In spite of the success and advanced technology of this process, there has been only little progress in the actual direct gap efficiency values.

The initial hope that the high efficiencies of GaAs diodes could be achieved for the entire direct gap region (ie up to 1·99 eV) has not been fulfilled for Ga(AsP). Close to the direct–indirect crossover, typical efficiency values are 0·2–0·6%, and the corresponding brightness value (related to current density) is typically 100 (ft L A^{-1} cm^2).

The general expression for the external quantum efficiency of direct transitions

$$\eta_{ext} = f\eta_i = f\left(1 + \frac{\tau_r}{\tau_{nr}}\right)^{-1}$$

formally implies that nonradiative recombination will shorten the lifetime τ_{nr} leading to a decrease in the internal efficiency η_i. The internal efficiency may also be reduced by a nonradiative current flow, for example a tunnel or surface leakage current. Most likely, the compositional grading of the vapour grown Ga(AsP) layer (typical thickness 25 μm) is still not sufficient to eliminate the strain effects due to lattice mismatch. Dislocations are generated in the graded transition layer, their number being proportional to the maximum composition gradient (Kishino *et al* 1972). Decorated dislocations can lead either directly or through their influence on the Zn-diffusion process to an efficiency decrease (Grenning and Herzog 1968, Stringfellow and Greene 1969). Furthermore, Ga vacancies or formation of Ga-vacancy–donor complexes can lead either to nonradiative transitions or to an undesirable infrared luminescence, which is quite frequently observed in common Ga(AsP) diodes around 1·3 eV (Stewart 1971). A general relationship for the influence of the disorder of atoms in the alloy systems on the efficiency has not been established so far.

The major area of recent progress in the Ga(AsP) system is in the transitions on the indirect gap side: significant improvements of the efficiency have been obtained by nitrogen doping (Groves *et al* 1971, Craford *et al* 1972, Craford *et al* 1973). In fact, the poor performance of the indirect transitions in the alloy systems had always been a contradiction of the good results obtained for GaP. Doping of Ga(AsP) with isoelectronic traps (N) gave a significant improvement. This is shown in figure 4, where brightness at a current density of 10 A cm^{-2} has been plotted as a function of composition. Without nitrogen, the curve shows the typical brightness maximum close to the direct–indirect transition (compare also figure 3). With nitrogen doping, the brightness is lowered on the direct gap (Γ) side, whereas it reaches similiar values as for GaP on the indirect (X) side. This is in complete accordance with the anticipated role of isoelectronic traps explained later for GaP. The maximum brightness (related to current density) for indirect transitions in Ga(AsP) has been reported to be 113 ft L A^{-1} cm^2 (Craford and Groves 1973). The emission spectra

B

change from a band-to-band or band–acceptor (Zn) emission on the direct side (Herzog *et al* 1969) to an unresolved NN pair luminescence on the indirect side. The shift to the NN pair emission, which has a slightly lower energy, means a slight reduction in brightness at the direct–indirect crossover. A resolution of the NN emission into individual lines attributable to the decay of excitons bound to isolated N atoms, nearest neighbour NN pairs, etc (N, NN_1, NN_3 etc) is only possible in the composition range near GaP ($x \geqslant 0\cdot 8$) (Craford *et al* 1972).

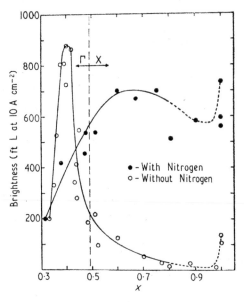

Figure 4. Brightness at $10\,\mathrm{A\,cm^{-2}}$ as a function of alloy composition for $GaAs_{1-x}\,P_x$ with and without nitrogen doping (Craford and Groves 1973).

Exactly at the transition point, direct band-to-band and NN transitions are simultaneously present leading to a 'resonant' enhancement of the total radiative transition probability. Interesting effects have been reported for this case in connection with stimulated emission (Holonyak *et al* 1972, Duke *et al* 1972).

Nitrogen doping of Ga(AsP) offers the possibility of manufacturing LEDs of reasonable brightness with variable colours ranging from red to green. Although the VPE process often does not offer optimum brightness figures, its attractiveness lies in its large scale production potential, in particular, in a very reproducible performance. These advantages have given the Ga(AsP) technology a lead at present over its main competitor, the GaP technology. The question is, whether this will remain so in the future.

2.4. Gallium phosphide LEDs

The two-component system GaP is basically easy to handle and has a good chance of becoming the leading technology for red and green/yellow LEDs in the future. Its main disadvantage is that it offers only these two colours and no continuous range; its advantage is higher efficiency and possibly a less complicated technology. The standard process for material synthesis has been liquid phase epitaxy (Lorenz and Pilkuhn 1966) on GaP substrates which means that no problems of lattice mismatch occur. The process, however, has shown some difficulties in connection with large scale production.

GaP has an indirect energy gap, and efficient radiative transitions at room temperature are obtained if isoelectronic traps are introduced either as atomic (N) or molecular traps (Zn–O). The recombination of excitons bound to these traps is the most favourable process for GaP LEDs.

The influence of isoelectronic traps in GaP is demonstrated in figure 5(*a*), which shows the dispersion relation for conduction and valence band. Different from the common donor impurity, the isoelectronic trap is characterized by a short range, square well potential, as depicted in figure 5(*b*). As a consequence, the wave function of the isoelectronic trap is fairly smeared in k-space. In particular, it contains contributions at $k=0$ (compare figure 5(*a*)) which are responsible for the high efficiency of the bound exciton recombination (compare Dean 1973b). In the case of the shallow isoelectronic trap (N), the room temperature efficiency is strongly limited, because the bound excitons are easily ionized thermally. The best values have been reported for LPE material where $\eta_{ext}=0\cdot7\%$ (Ladany and Kressel 1972, Logan *et al* 1971). A good efficiency at room temperature is obtained for a typical trap concentration of 3×10^{18} cm^{-3}. Higher N concentrations ($\simeq 10^{20}$ cm^{-3}) can be obtained in VPE material; however, here the efficiencies are very much lower with $\eta_{ext}=0\cdot12\%$ (Hart 1973). The green emission of GaP has the advantage (common with the red to green emission from Ga(AsP) alloys) that no saturation is observed with increasing current density through the diode.

In VPE GaP, where nitrogen concentrations of more than 10^{20} cm^{-3} can be incorporated, the emission peak shifts to longer wavelengths, and a maximum efficiency was actually observed for yellow emission (Nicklin *et al* 1971). It is possible in fact to adjust the emission wavelength within a narrow spectral range. Another possibility for a yellow GaP LED is the combination of red and green emission (n-GaP : N, p-GaP : Zn–O) (Rosenzweig *et al* 1971). Finally, doping with Mg–O leads to a yellow emission (Bhargava *et al* 1972) whose efficiency, however, is not yet satisfactory.

Adjustment of the wavelength of the emission in the direction of shorter wavelengths is possible by addition of Al to GaP. Recently, green emission at 5350 Å was investigated in the system (AlGa)P (Sonomura *et al* 1973).

The decay of the exciton bound to the Zn–O molecular isoelectronic trap

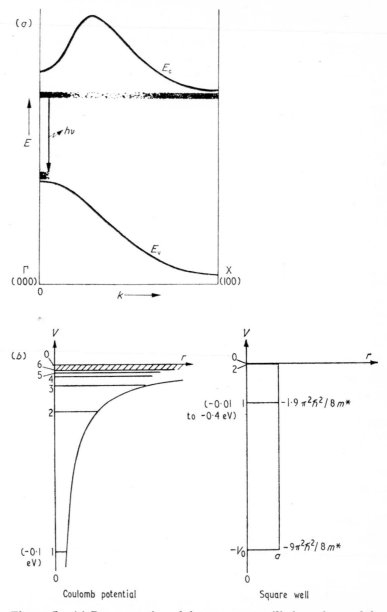

Figure 5. (*a*) Representation of the wave vector (*k*) dependence of the probability densities of an electron bound to an isoelectronic trap (Thomas 1971). (*b*) Potential wells and associated bound states for Coulomb binding and for binding in a three-dimensional square well (depth V_0, radius *a*). Binding energies for coulombic centres and isoelectronic traps for indirect gap semiconductors are indicated (Dean 1973b).

has the highest efficiency of all room temperature recombination processes in GaP. The reason is that thermal dissociation of the exciton bound to Zn–O hardly takes place at all at temperatures as high as room temperature. The highest efficiency values published for red-emitting LEDs lie little over 12% (Solomon and DeFevere 1972, K Maeda *et al* 1973 private communication). The highest oxygen concentration incorporated into LPE GaP is only about 10^{16} cm^{-3} (Henry *et al* 1973), and as a consequence the red emission saturates quickly with increasing diode current density. This is demonstrated in figure 6,

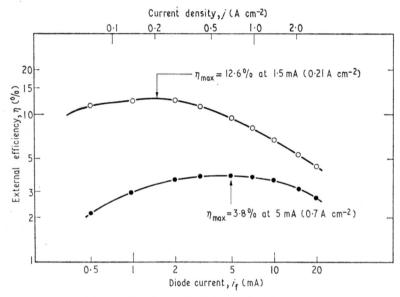

Figure 6. External efficiency of GaP: Zn–O diodes at room temperature as a function of diode current (K Maeda *et al* 1973 private communication).

where the efficiencies of an 'average' diode and of an 'excellent' diode are plotted as a function of current density. In most cases, saturation effects start at current densities of $\geqslant 10$ A cm^{-2}. The highly efficient GaP red LEDs can be driven at currents in the order of milliamperes which means that the general drawback of a high power consumption, attached generally to LEDs, disappears.

Progress in the further improvement of the internal efficiencies of GaP and other semiconductors will come through the understanding and control of nonradiative centres. Although nonradiative transitions have been discussed for many years, little progress can be reported. The Auger effect is certainly a good candidate for nonradiative recombination, and three possible Auger transitions in GaP are presented in figure 7. In process (*a*), during the decay of an exciton bound to a neutral donor (S), the energy is transferred to the donor

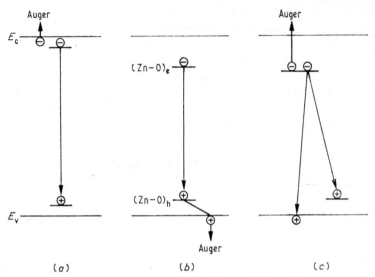

Figure 7. Schematic representation of three different Auger processes. (*a*) Auger process associated with excitonic decay at a neutral donor. (*b*) Auger process associated with a bound exciton decay at a molecular isoelectronic trap (Zn–O) with a free hole as a third particle. (*c*) Auger recombination at a doubly ionizable donor (Bhargava 1972).

electron which is lifted into the conduction band and which equilibrates subsequently under emission of phonons. In process (*b*), that is the bound exciton decay at a Zn–O molecular trap, the energy can be transferred to a free hole. This is a process for which a high concentration of free holes is required. In process (*c*), the recombination takes place at a doubly ionizable donor, and the energy is transferred to the second donor electron. In processes (*a*) and (*c*), the Auger particle is always present at the site of the recombination which is an attractive feature. Possibly, the complex Si_{Ga}–O_P is a double deep donor in GaP of the type needed for process (*c*).

In general, the nature of nonradiative centres in GaP (which still take at least 40% of the recombination current) has not been identified. There has been progress, on the other hand, in accumulating more quantitative information on centres in GaP by new experimental techniques like photocapacitance and thermally stimulated current measurements, infrared spectroscopy and activation analysis. A surprising number of new centres was found in GaP grown by the 'liquid encapsulation Czochralski' (LEC) method, the standard substrate material for GaP epitaxy. Figure 8 represents recent measurements of thermally stimulated currents of undoped LEC GaP (figure 7(*a*)) and n-type LEC material (figure 7(*b*)). A prominent level is found at 0·36 eV in many n-samples and is associated with the incorporation of common donors (Te, Se, S). Altogether, up to nine deep donor states (ranging from 0·27 eV to 0·90 eV) have

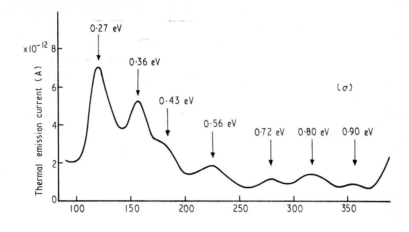

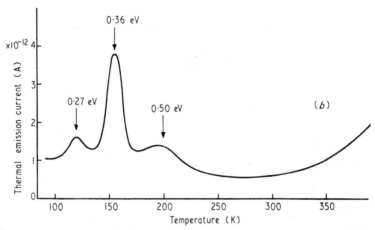

Figure 8. Thermally stimulated current spectra of n-type LEC GaP. (a) Undoped crystal $N_d - N_a \simeq 10^{17}$ cm^{-3}. (b) Te-doped crystal $N_d - N_a \simeq 4 \times 10^{17}$ cm^{-3} (Fabre *et al* 1973).

been seen which makes it very clear why the quantum efficiency of LEC GaP is extremely low.

Better quality bulk GaP will be one of the technological challenges in the future. It is desirable to use bulk crystals immediately for diode fabrication and not as substrates for epitaxy only. A promising approach is the method called 'synthesis, solute diffusion' (SSD) (Kaneko *et al* 1973, Okuno *et al* 1973) as investigated for example at the Sony Research Center and its principle is explained in figure 9: phosphorus evaporating from the lower part of the vessel reacts with the Ga melt and forms a GaP film. Phosphorus diffusion through

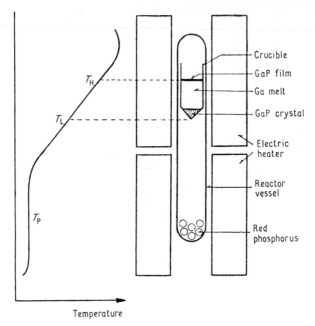

Temperature

Figure 9. Schematic diagram of an apparatus for GaP synthesis by the SSD method (Kaneko *et al* 1973).

the Ga melt leads to crystal growth at the bottom of the crucible. SSD crystals have greatly improved luminescence properties; LEDs can be made success-fully by a true one-step LPE process (instead of a two-step LPE process on LEC substrates). Red LED efficiencies of 7·5% and green LED efficiencies of 0·15% have been obtained for SSD material (Kaneko *et al* 1973).

On the other hand, crystal sizes are not satisfactory in view of the dramatic improvements obtained for LEC GaP (LEC crystal diameters have reached 60 mm, total crystal weight 450 g).

The major progress in the development of GaP LEDs will come through the materials scientist: development of better and cheaper bulk material, improve-ment of epitaxy processes and control of nonradiative recombination are the major tasks.

Considerable cost reductions together with an increasing demand for LEDs (both Ga(AsP) and GaP) are predicted for the coming years. One of the predic-tions (Craford and Groves 1973) concerning the selling price of optoelectronic devices in comparison with Si integrated circuits is seen in figure 10 and does not require comment. In comparing the development of the two technologies, one should keep in mind that microminiaturization in optoelectronic displays will not play the same role as in the silicon technology, for obvious reasons. The rapid rate of development of the existing major optoelectronic display

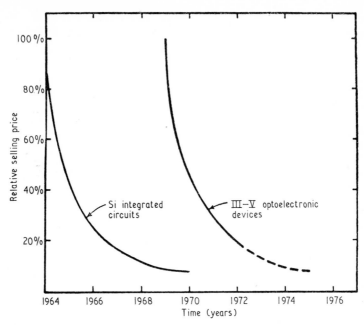

Figure 10. Price development of Si integrated circuits and a combination of optoelectronic devices (Craford and Groves 1973).

technologies (Ga(AsP) and GaP) and the anticipated price developments will make it very difficult for any other semiconductor to compete in this field.

3. Optical communication

Another modern field of optoelectronics deserving special attention is 'optical communication' because of the many expectations for future applications.

The following sequence of devices is needed for optical communication: lasers $\rightarrow$ modulators $\rightarrow$ couplers $\rightarrow$ transmitters $\rightarrow$ repeaters $\rightarrow$ detectors. Spontaneous emitters (eg GaAs LEDs) can also serve in optical communication, but the field has certainly lost its attractiveness without the use of coherent laser radiation.

Some general points which should be considered for a development of optical communication are discussed first.

(i) SIC compatibility of the electro-optic devices. Si devices will have to be added at the beginning of the device line (in connection with lasers and modulators) and at the end of the device line. Most likely, the detector itself will be of silicon.

(ii) Agreement upon a common wavelength of optimum performance for all devices involved. Where Si is the detector material, the long wavelength limit is determined by the energy gap of Si.   The wavelength of the lasers with the highest technological potential (GaAs injection or Nd ultraphosphate lasers) together with the optimum wavelength for optical fibres and modulators suggests that the best common wavelength for optical communication will lie between 0·7 and 1·1 μm.

(iii) Lifetime, reliability, small size and ruggedness of devices are conditions which favour a solid-state technology.

(iv) Integrated optics.   The optoelectronic components should be able to be incorporated into integrated optics, which will probably play an important role for optical communication in the more distant future.

3.1. *Transmission through optical fibres*

Spectacular progress has been achieved for the transmission characteristics of low loss optical fibres.   Figure 11(a) shows the attenuation of a low-loss multi-mode glass fibre as a function of wavelength.   Waveguiding is obtained through a suitable variation of the refractive index across the diameter of the fibre.   The lowest loss of 7 dB km^{-1} is reached in figure 11(a) at a wavelength of 1·07 μm; another low loss region lies between 0·8 and 0·9 μm.   This coincides nicely with the emission wavelength of GaAs lasers, whereas Nd or (InGa)As lasers would fall into the first attenuation minimum.   The lowest published attenuation values for multimode glass waveguides are 4 dB km^{-1} at 0·8–0·85 μm and at 1·05 μm (Keck *et al* 1973), while unofficial reports quote even lower values.

The question of the ultimate limits of the attenuation can be discussed in connection with figure 11(b), where the attenuation is split up into absorptive and scattering components.   Rayleigh scattering (broken curve) with a λ^{-4} dependence is one principal limitation.   The remaining experimental absorption curve (dotted curve) can be explained at wavelengths between 0·65 and 1·6 μm by OH absorption (full curve).   There is hope that the absorptive attenuation can be removed or at least reduced to values below 1 dB km^{-1} in the wave-length range between 0·7 and 1·1 μm.   Intrinsic glass scattering then is the only limitation, and attenuation values as low as 1–2 dB km^{-1} should be achieved beyond 0·8 μm.

3.2. *GaAs lasers: recent progress*

For optical communication, the room temperature lasing wavelength of GaAs injection lasers (near 0·9 μm) appears to be quite convenient.   If desired, the emission can be shifted to shorter wavelength (eg 0·8–0·85 μm) by the addition of Al ((GaAl)As lasers), or to longer wavelength (eg 1·05 μm) by the addition of In ((InGa)As lasers) in order to match the attenuation minima in the glass fibre characteristics (figure 11(a)).   However, the technology of ternary alloy lasers like (InGa)As is more difficult.

High efficiency, small size, mechanical ruggedness, easy sic-compatible driving, and the possibility of direct modulation at rates up to Gbit/s (Chown

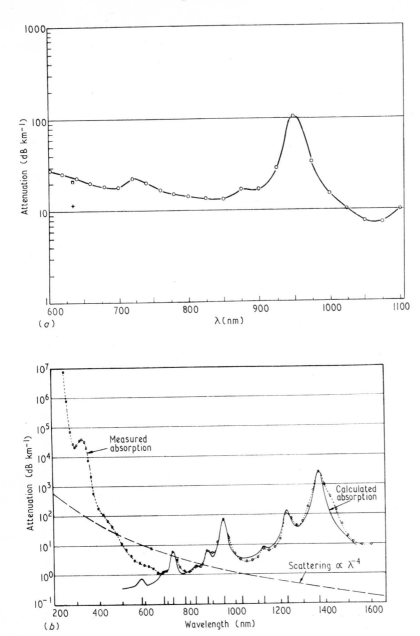

Figure 11. (*a*) Attenuation as a function of wavelength for a low loss multimode glass fibre (Keck *et al* 1972). (*b*) Measured and calculated absorptive and scattering attenuations as a function of wavelength. To obtain the total attenuation, scattering and absorptive components have to be added (Keck *et al* 1973).

et al 1973) have been the attractive features of injection lasers, together with the capability of low cost production.   On the other hand, its weak points have been (i) high room temperature thresholds, (ii) filamentary lasing, and (iii) degradation during continuous operation.   Recent progress achieved in these areas will be reviewed in this paper.   Other points which require attention are mode control, instabilities and high power performance.

The 'direct' band-to-band and the band-to-acceptor recombination are the common laser processes in highly doped GaAs.   Their quantitative description is complicated by the fact that impurity states mix with band states giving rise to 'tails', which modify the density of states functions describing the gain (Stern 1966, 1973, Hwang 1970).   Furthermore, it was found that recombination in highly doped GaAs takes place without k-selection rules (Lasher and Stern 1964) which is not surprising considering the participation of impurity states.   Recently it was found, however, that even in high purity GaAs, no k-selection rules are required to describe the low temperature gain (Göbel *et al* 1973).   The high temperature gain has not been analysed sufficiently well in order to decide on the issue of k-selection rules which, for practical laser devices, might have an appreciable influence on the temperature dependence of the threshold current density j_t.

The rapid increase of j_t with temperature has been the main obstacle for room temperature continuous operation: early values for j_t (300 K) for diffused homojunction lasers have been as high as 10^5 A cm^{-2}.   Theoretically, the

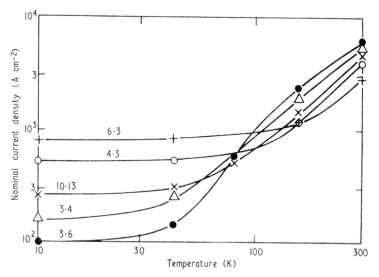

Figure 12.   Nominal current density required to reach a gain of 50 cm^{-1} for two n samples and three p samples as a function of temperature. Integers labelling each curve give the donor and acceptor concentration, respectively, in units of 10^{18} cm^{-3} (Stern 1973).

presence of k-selection rules leads to a flatter temperature dependence of j_t (Hall 1963, Pikus 1966), but in practice it is very doubtful whether this approach to the problem should be followed by using high purity or low doped GaAs. In fact, the opposite approach, namely to use highly doped and compensated active layers, has been quite successful, because of the influence of tail states on the temperature dependence of j_t (Stern 1966, 1973).   This is demonstrated in figure 12 where a measure of the threshold current density (nominal current

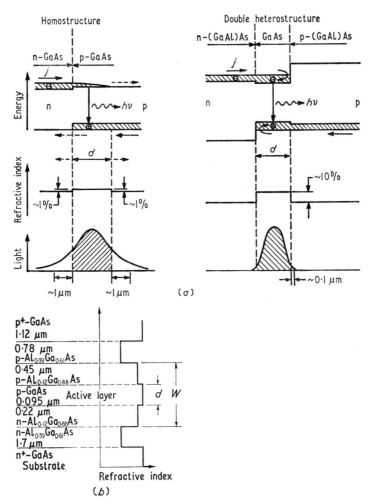

Figure 13. (a) Schematic diagram of band edges for forward bias, refractive index and vertical light distribution for homo and double heterostructure lasers (Hayashi *et al* 1971).   (b) Schematic diagram of a separate confinement heterostructure (SCH) laser (Panish *et al* 1973).

density to reach a gain of 50 cm⁻¹) has been plotted against temperature. At room temperature, higher doped and more compensated samples have a lower threshold current density for both p and n material, and figure 12 suggests that compensated n-layers (eg $N_d = 6 \times 10^{18}$, $N_a = 3 \times 10^{18}$ cm⁻³) should give the best results (Stern 1973).

High Si or Ge doping has successfully been used for the active layers of injection lasers and recently low doping levels have been also found to give quite good results (Hwang and Dyment 1973). The question of the optimum doping level for the lowest threshold seems to have lost its importance in view of other influences like heterostructure parameters.

The most dramatic improvements leading to continuous operation at room temperature have been achieved by the development of double heterostructure (DH) lasers (Hayashi *et al* 1970). In figure 13(*a*), the band edges for forward bias, the refractive index change and the vertical light distribution are compared between homostructures and heterostructures (Hayashi *et al* 1971). The additional energy barrier in the heterostructure leads to an *electrical confinement* of the injected electrons and holes (improving the laser gain coefficient), and the large refractive index change leads to an *optical confinement* (reducing the losses due to reabsorption of light in the passive layers). The structures are conveniently fabricated by LPE growth of successive GaAs and (GaAl)As layers.

The threshold current density can be lowered significantly if the thickness of the active layer, *d*, is reduced. This is shown in figure 14 for the case of an undoped active layer which was investigated recently (Nakada *et al* 1973).

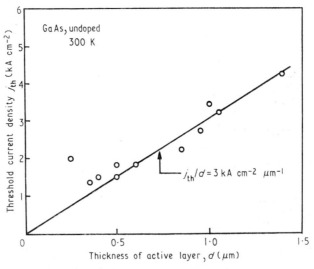

Figure 14. Threshold current density as a function of active layer thickness, *d*, for double heterostructure lasers with undoped active layers (Nakada *et al* 1973).

Threshold current densities of $\geqslant 1$ kA cm^{-2} can be obtained in this way, and continuous operation of these devices at room temperature is possible.

A further improvement was obtained by an additional sophistication of LPE technology: heterostructures with separate electrical and optical confinement (SCH lasers) (Panish *et al* 1973, Thompson and Kirkby 1973). A schematic diagram of this separate confinement heterostructure laser with typical dimensions of the epitaxial layers is depicted in figure 13(*b*). Electrons and holes are confined in a very narrow layer of width *d* and the light is confined (through an additional large refractive index change) in the layer of width *W*. Threshold current densities as low as 575 A cm^{-2} have been published for unsilvered Fabry–Pérot units (Thompson and Kirkby 1973); unofficial reports mention values of 300 A cm^{-2}.

The production technology for SCH lasers has reached a very high standard: fabrication of 5 or 7 layer structures with layer thicknesses down to 0·04 μm and the controlled incorporation of impurities under these conditions is quite a respectable achievement. The proper control of a multilayer technology allows a rather optimistic outlook on future laser performance as continuous operation at current levels in the order of 10 mA and external efficiencies as high as 50 to 70% seem possible. Furthermore, double heterostructure lasers with narrow active layers are promising devices for high power generation: peak powers of over 40 W/mm junction width at angular beam spreads as low as 16° have been demonstrated to be feasible (Kirkby and Thompson 1973).

One of the conditions for the application of DH lasers in optical communication is a reasonable far field pattern and a good mode control. This is not possible in the case of filamentary laser action which normally occurs in wide junction lasers. A solution to the problem of getting a uniform lasing region with a well defined far field pattern is the introduction of the 'stripe geometry structure'. It was first made by providing a thin stripe contact (with oxide masking) to the p$^+$ side of the laser structure leading to a 'Hermite–Gaussian' intensity pattern for transverse modes (Dyment 1967). The proper construction of a stripe laser geometry has become an important engineering task, because it determines the optical as well as thermal characteristics of the device (D'Asaro 1973). Control of heat conduction in DH stripe geometry structures has become essential for CW or high duty cycle performance.

Figure 15(*a*) and (*b*) show two possible techniques for a stripe geometry configuration: in the proton bombardment technique of figure 15(*a*) (Dyment *et al* 1972), high resistivity regions outside the stripe are introduced preventing a current flow. Optical losses in the bombarded regions can be reduced by subsequent annealing. The proton bombardment technique is a very fast method for device fabrication.

A mesa stripe structure (Tsukada *et al* 1972) is shown in figure 15(*b*). It should in principle have the best optical and electrical confinement, and very low threshold currents can be achieved with this structure.

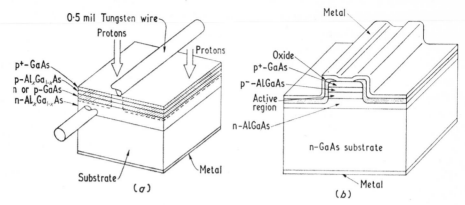

Figure 15. (*a*) Stripe geometry laser structure produced by proton bombardment (Dyment *et al* 1972). (*b*) Mesa stripe laser structure (Tsukada *et al* 1972).

In more recent work, internally striped planar (ISP) lasers have been made by Zn-diffusion techniques, and in figure 16 three possibilities are compared. In each case, the Zn-diffusion (hatched area) prevents or reduces the current flow outside the stripe. In case (*a*), the major current flow will be across the p-GaAs, n-(GaAl)As heterojunction (smaller barrier) instead of the (GaAl)As p–n junction with a larger barrier (Susaki *et al* 1973). In the other two configurations in figure 16, a reverse biased p–n junction in the n-(GaAl)As or n-GaAs layer prevents a current spreading away from the stripe (Yonezu *et al*

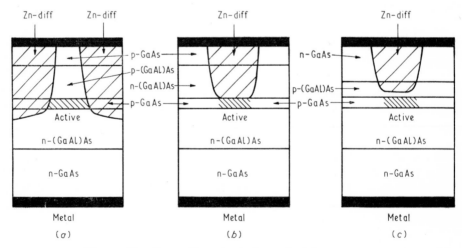

Figure 16. Internally striped planar (ISP) laser structure produced by Zn-diffusion.

1973, Takusagawa *et al* 1973). All these diffused ISP laser structures should have very good thermal characteristics, it should be possible to apply photo-resist techniques, and one should obtain a very small stripe width.

The greatest obstacle in the application of injection lasers to optical communication has been their rapid degradation during continuous operation. One has to distinguish between (i) catastrophic self-damage at extremely high power levels and (ii) slow degradation at low power levels (see Eliseev 1973 for a review of work on degradation up to September 1972).

During the slow degradation, the light power output decreases steadily with time of operation. Both an increase in threshold current density and a decrease in the quantum efficiency can be made responsible for the slow degradation. Until very recently, this problem appeared to be quite formidable, because no experimental method of avoiding or reducing the degradation was known, and no agreement on one single degradation mechanism could be reached. It was generally clear that diffusion and/or generation of defects should be involved.

This situation has improved significantly with the application of the following new experimental techniques, which all indicate that local strain and formation of dislocation networks is causing the slow degradation:

(i) Investigation of strain induced birefringence (photoelastic effect) (Hartman and Hartman 1973).

(ii) Photoluminescence topography (Nakada *et al* 1973, Johnston and Miller 1973), cathodoluminescence topography (Yonezu *et al* 1973, Nannichi and Hayashi 1973 private communication), electroluminescence topography (De Loach *et al* 1973).

(iii) Transmission electron microscopy (Petroff and Hartman 1973).

A definite relation between strain, probably introduced as bonding strain, and the GaAs laser lifetime was found in birefringence experiments (Hartman and Hartman 1973). The various topography investigations revealed dark, non-luminescent lines (stripes, regions) in the junction plane, which appeared in connection with laser degradation. On the other hand, these lines were found to be related to local strain. The dark lines had been discovered earlier in degraded GaAs LEDs (Baird *et al* 1967, Mettler and Pawlik 1972) and their effect seems to be also present and much enhanced in lasers. Nonradiative recombination centres have formed in these dark lines leading to a reduction in the gain as well as an increase in the losses. A local temperature rise in the regions with larger absorption can accelerate the degradation. The mechanism of degradation in the strained regions will involve formation of dislocations, enhanced impurity migration and decoration of dislocations (Kressel 1974, Petroff and Hartman 1973).

The answer to the degradation problem is presently the development of technologies eliminating strain. Large improvements of laser lifetime have been obtained by applying a low strain bonding technique (Hartman *et al* 1973),

c

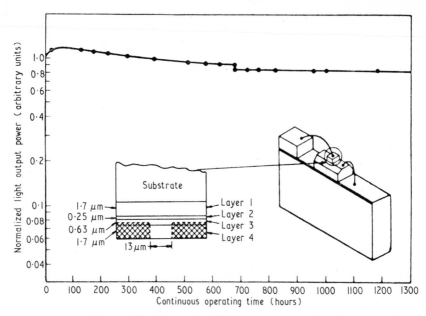

Figure 17. Normalized light output as a function of continuous operating time at 30 °C (nitrogen ambient) (Hartman *et al* 1973).

or by special doping (eg AlP) of the active layer (Y Nannichi and I Hayashi 1973 private communication, Rozgonyi and Panish 1973). Figure 17 shows a published lifetime test of a DH laser showing only little change of the light output during 1300 hours of CW operation. Much larger lifetimes, exceeding 4000 hours, have been reached in the meantime (unpublished results), and lifetimes of 10^4 hours should be obtainable in the near future.

After the elimination of strain effects and dislocation networks, no other degradation mechanism can be seen today that limits the laser lifetime.

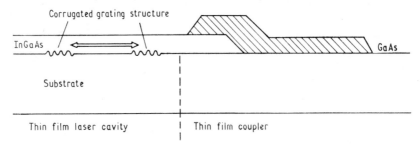

Figure 18. Schematic diagram of thin film laser, coupler and dielectric waveguide in integrated optics.

3.3. Integrated optics

A detailed discussion of the role of integrated optics in optical communication is beyond the scope of this paper. Integrated optics require the incorporation of elements like lasers, modulators etc in connection with thin film waveguides, thin film couplers etc. An example of how one of these elements, a semiconductor laser (eg (InGa)As), has to be engineered in integrated optics is demonstrated in figure 18. Instead of laser mirrors, a corrugated grating structure selecting a laser wavelength throughout its Bragg condition has to be used. The laser emission can then be coupled by a special thin film coupler into a thin film waveguide (eg a GaAs film). Recently, a GaAs laser of the type similar to that shown in figure 18 has been produced (Nakamura *et al* 1973). A schematic

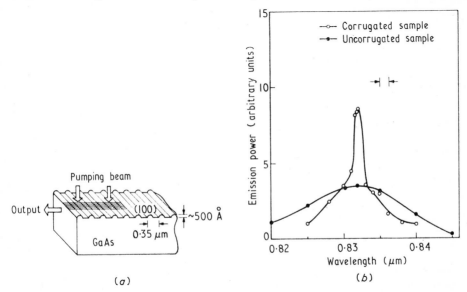

Figure 19. (*a*) Schematic structure of an optically pumped GaAs surface laser. (*b*) Emission spectrum of a corrugated and an uncorrugated sample. Pumping intensity is $5 \times 10^5 \, \mathrm{W \, cm^{-2}}$ (Nakamura *et al* 1973).

structure of the optically pumped GaAs surface laser is shown in figure 19(*a*), and the effect of the corrugated structure on the emission spectra in figure 19(*b*). A strongly narrowed stimulated emission line is found for the corrugated laser structure.

Although these first results appear to be very modest, the attractiveness and potential of integrated optics in optical communication will stimulate further development. This field, however, together with advanced technologies like ion implantation and molecular beam epitaxy, will have their chances in the more distant rather than in the near future.

References

Archer R J 1972 *J. Electron. Mater.* **1** 128
Ashley K L and Strack H A 1969 *Proc. 2nd Int. Symp. on GaAs* (London: Institute of Physics) p123
Baird J R, Pittman G E and Leezer J F 1967 *Proc. 1st Int. Symp. on GaAs* (London: Institute of Physics) p113
Bhargava R N 1972 *Philips Tech. Rev.* **32** 261
Bhargava R N *et al* 1972 *Appl. Phys. Lett.* **20** 227
Chown M, Goodwin A R, Lovelace D F, Thompson G H and Selway P R 1973 *Electron. Lett.* **9** 34
Craford M G and Groves W O 1973 *Proc. IEEE* **61** 862
Craford M G, Keune D L, Groves W O and Herzog A H 1973 *J. Electron. Mater.* **2** 137
Craford M G, Shaw R W, Groves W O and Herzog A H 1972 *J. Appl. Phys.* **43** 4075
D'Asaro L A 1973 *J. Luminescence* **7** 310
Dean P J 1973a *Int. Conf. Solid State Devices, Tokyo*
—— 1973b *J. Luminescence* **7** 51
De Loach B C, Hakki B W, Hartman R L and D'Asaro L A 1973 *Proc. IEEE* **61** 1042
Duke C B *et al* 1972 *J. Appl. Phys.* **43** 5134
Dyment J C 1967 *Appl. Phys. Lett.* **10** 84
Dyment J C, D'Asaro L A, North J C, Miller B I and Ripper J E 1972 *Proc. IEEE Lett.* **60** 726
Eliseev P G 1973 *J. Luminescence* **7** 338
Fabre E, Bhargava R N and Zwicker W K 1973 to be published
Göbel E, Herzog H, Pilkuhn M H and Zschauer K H 1973 *Solid St. Commun.* **13** 719
Grenning D A and Herzog A H 1968 *J. Appl. Phys.* **39** 2783
Groves W O, Herzog A H and Craford M G 1971 *Appl. Phys. Lett.* **19** 184
Hakki B W 1971 *J. Electrochem. Soc.* **118** 1469
Hall R N 1963 *Solid St. Electron.* **6** 405
Hart P B 1973 *Proc. IEEE* **61** 880
Hartman R L, Dyment J C, Hwang C J and Kuhn M 1973 *Appl. Phys. Lett.* **23** 181
Hartman R L and Hartman A R 1973 *Appl. Phys. Lett.* **23** 147
Hayashi I, Panish M B, Foy P W and Sumski S 1970 *Appl. Phys. Lett.* **17** 109
Hayashi I, Panish M B and Reinhart F K 1971 *J. Appl. Phys.* **42** 1929
Henry C H, Kukimoto H, Miller G L and Merritt F R 1973 *Phys. Rev.* **B7** 2499, 2486
Herzog A H, Groves W O and Craford M G 1969 *J. Appl. Phys.* **40** 1830
Holonyak Jr N, Dupuis R D, Macksey H M, Craford M G and Groves W O 1972 *J. Appl. Phys.* **43** 4148
van Houten S 1973 *Proc. IEEE* July
—— 1974 this volume p131
Hwang C J 1970 *Phys. Rev.* **B2** 4117, 4126
Hwang C J and Dyment J C 1973 *J. Appl. Phys.* **44** 3240
Johnston W D and Miller B I 1973 *Appl. Phys. Lett.* **23** 192
Kaneko K *et al* 1973 *Proc. IEEE* **61** 884
Keck D B, Maurer R D and Schultz P C 1973 *Appl. Phys. Lett.* **22** 307
Keck D B, Schultz P C and Zimar F 1972 *Appl. Phys. Lett.* **21** 215
Kirkby P A and Thompson G H 1973 *Appl. Phys. Lett.* **22** 638
Kishino S, Ogirima M and Kurata K 1972 *J. Electrochem. Soc.* **119** 617
Kressel H 1974 *J. Phys., Paris* to be published
Kressel H, Hawrylo F Z and Almeleh N 1969 *J. Appl. Phys.* **40** 2248
Kressel H, Nuese C J and Ladany I 1973 *J. Appl. Phys.* **44** 3266

Ladany I and Kressel H 1972 *Proc. IEEE Lett.* **60** 1101
Lasher G J and Stern F 1964 *Phys. Rev.* **133** A553
Loebner E 1973 *Proc. IEEE* **61** 837
Logan R A, White H G and Wiedmann W 1971 *Solid St. Electron.* **14** 55
Lorenz M R and Pilkuhn M H 1966 *J. Appl. Phys.* **37** 4094
Maruska H P and Pankove J I 1967 *Solid St. Electron.* **10** 917
Mettler K and Pawlik D 1972 *Siemens Forschg. Entwickl. Berichte* **1** 274
Nakada O, Ito R, Nakashina H and Chinone N 1973 *Int. Conf. on Solid State Devices, Tokyo*
Nakamura M, Yariv A, Yen H W, Somekh S and Garwin H L 1973 *Appl. Phys. Lett.* **22** 515
Nicklin R, Mobsby C D, Lidgard G and Hart P B 1971 *J. Phys.* C: *Solid St. Phys.* **4** 344
Nuese C J, Sigai A G, Abrahams M S and Gannon J J 1973 *J. Electrochem. Soc.* **120** 956
Okuno Y, Suto K and Nishizawa J 1973 *J. Appl. Phys.* **44** 832
Onton A and Lorenz M R 1971 *Proc. 3rd Int. Symp. on GaAs* (London: Institute of Physics) p222
Onton A, Lorenz M R and Reuter W 1971 *J. Appl. Phys.* **42** 3420
Panish M B, Casey H C, Sumski S and Foy P W 1973 *Appl. Phys. Lett.* **22** 590
Pankove J I 1973 *J. Luminescence* **7** 114
Petroff P and Hartman R L 1973 *Appl. Phys. Lett.* **23** 469
Pikus G E 1966 *Sov. Phys.–Solid St.* **7** 2854
Pilkuhn M H and Rupprecht H 1965 *J. Appl. Phys.* **36** 684
Rosenzweig W, Logan R A and Wiegman W 1971 *Solid St. Electron.* **14** 655
Rozgonyi G A and Panish M B 1973 *Appl. Phys. Lett.* **23** 533
Shay J L 1972 *Proc. 11th Int. Conf. on Phys. of Semicond., Warsaw* (Amsterdam: Elsevier) p787
Sigai A G, Nuese C J, Enstrom R E and Zamerowski T 1973 *J. Electrochem. Soc.* **120** 947
Solomon R and DeFevere D 1972 *Appl. Phys. Lett.* **21** 257
Sonomura H, Nanmori T and Miyauchi T 1973 *Appl. Phys. Lett.* **22** 532
Stern F 1966 *Phys. Rev.* **148** 186
—— 1973 *IEEE J. Quantum Electron.* **QE9** 290
Stewart C E E 1971 *J. Crystal Growth* **8** 259
Stringfellow G B and Greene P E 1969 *J. Appl. Phys.* **40** 502
Susaki W, Namizaki H, Kan H and Ito A 1973 *J. Appl. Phys.* in the press
Takusagawa M, Ohsaka S and Takagi N 1973 to be published
Thomas D G 1971 *IEEE Trans. Electron. Dev.* **ED18** 621
Thompson G H and Kirkby P A 1973 *Electron. Lett.* **9** 295
Triebwasser S 1973 *Int. Conf. Solid State Devices, Tokyo*
Tsukada *et al* 1972 *Appl. Phys. Lett.* **20** 344
Woodall J M, Rupprecht H and Reuter W 1969 *J. Electrochem. Soc.* **116** 899
Yonezu H *et al* 1973 *Int. Conf. Solid State Devices, Tokyo*

Solid-state devices in watches

F Leuenberger

Centre Electronique Horloger SA, Rue A-L Breguet 2,
Neuchâtel, Switzerland

Abstract

Existing micropower integrated circuits enable the designer to implement logical functions beyond the particular function of frequency division. A brief description of a logically adjustable frequency divider introduces the subject.

The first part deals with the realization and performance of silicon gate CMOS circuits. Problems related to complementary dynamic MOS circuits are emphasized. The description of a concept that enables the simultaneous integration of bipolar pnp and npn transistors with p-channel and n-channel MOSTs rounds out the picture.

The second part gives a general review of electronic displays which are potentially applicable to watches. Two examples are given for the integration of decoding and display functions in monolithic and hybrid form respectively. In particular, we describe an ambient powered analog electroluminescent diode display system. The utilization of a solar battery as the main energy source for the display makes feasible an active display whose contrast is independent of the changing ambient light level.

1. Introduction

In this review we are concerned with all solid-state watches and we deliberately omit hybrid electronic–electromechanical solutions (Forrer 1972) from the discussion. In figure 1 we have shown the building blocks of a generalized watch system. To avoid getting lost in superficialities we shall further concentrate on the building blocks 'Logic' and 'Display'. These two blocks represent the bulk of the electronics in an all solid-state watch anyway. Only

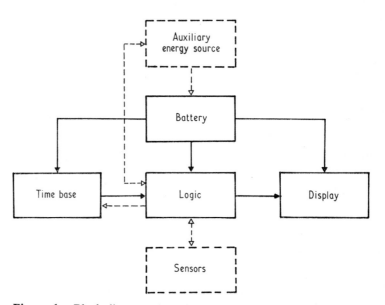

Figure 1. Block diagram of an electronic watch.

near the end, in connection with a system using an electroluminescent display shall we have to say a few words about auxiliary power sources and sensors.

The utilization of advanced micropower integrated circuits allows the realization of logic functions beyond the particular function of frequency division. Let us illustrate this point by means of an example. In a commercially available quartz watch, the oscillator frequency is usually adjusted in three successive steps.

(i) Coarse adjustment to $\pm 2 \times 10^{-5}$ by grinding the quartz.

(ii) Fine adjustment by evaporating a small mass on the mounted quartz prior to final encapsulation.

(iii) The final adjustment is made electrically by means of a small variable capacitor.

Points (i) and (ii) contribute sensibly to the cost and the addition of a trimmer to displace the oscillator frequency slightly from its natural resonant frequency degrades to some degree the inherent stability of the oscillator.

The stability of the oscillator is no doubt of paramount importance. The accuracy of the oscillator, on the other hand, is of no importance. In fact, it is the accuracy of the frequency of the output signal driving the display that must be ascertained, and this can very well be implemented by logical means (Vittoz *et al* 1971). In such a system the quartz is only coarsely adjusted and the trimmer is omitted. In this way it is possible to reduce the cost of the quartz and at the same time maintain its inherent stability. The desired accuracy of the output signal frequency is obtained by using a logically adjustable frequency divider in connection with an alterable read only memory (ROM). An n-bit memory will provide 2^n successive intervals of adjustment. The number of intervals and thus the total range of adjustment may be varied to whatever extent one may wish, and this without any degradation of the stability. In the simplest case, the alterable ROM may consist of a row of electromechanical switches. In a more sophisticated approach one might use nonvolatile alterable ROMs such as, for instance, the MNOS or the FAMOS. Figure 2 shows the block diagram of the

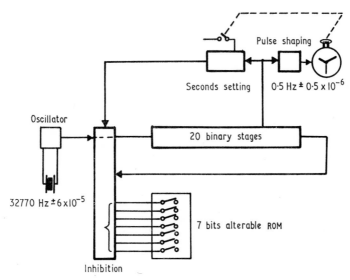

Figure 2. Block diagram of quartz watch incorporating a logically adjustable frequency divider.

system. The 20-stage frequency divider is preceded by an inhibiting circuit. During each period of the frequency divider output signal, this circuit eliminates, one after the other, a certain fraction of the pulses delivered by the oscillator. Each pulse eliminated corresponds to a reduction of $(0.5)^{20}$, or about 1 µs, of the output signal period. A seven-bit memory, for instance, can eliminate up to 127 pulses per cycle of adjustment, corresponding to $\pm 6 \times 10^{-5}$ for the complete range of adjustment.

2. Technology of micropower integrated circuits

Modern bipolar circuits where high value resistors are replaced by active devices are very useful in systems with oscillator frequencies up to 32 kHz (Thommen and Ruegg 1971). Complementary MOS circuits on the other hand provide micropower operation at low supply voltages at several MHz. Let us turn now to the description of a fabrication sequence for silicon gate CMOS structures (Vittoz *et al* 1972) (figure 3). Starting material is n-type (100) orientation silicon with a resistivity of 4 ± 1 ohm cm. The p-type wells are made either by

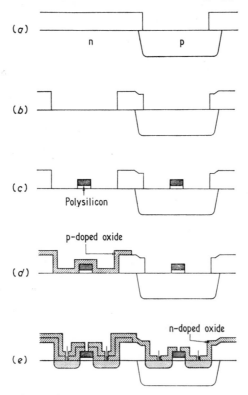

Figure 3. Fabrication sequence for complementary silicon gate MOS circuits.

a sealed capsule diffusion or else by boron implantation, followed by a drive-in step. The target value for the final boron surface concentration is about 8×10^{15} cm^{-3}. Windows are now opened delineating source, drain and gate regions for both the p-channel and n-channel devices. The 0·1 μm thick gate oxide is now grown in dry oxygen followed by a 15 minutes *in situ* annealing in He at the oxidation temperature. A 0·6 μm thick layer of polycrystalline silicon is now deposited from the vapour phase. This is followed by selective etching of the polycrystalline silicon and removal of the gate oxide in the source and drain regions.

Next, a boron-doped oxide is deposited over the entire wafer surface and then selectively removed over the p-type regions. A phosphorus-doped oxide is deposited, again over the entire wafer surface, and source as well as drain regions of both p-channel and n-channel devices are diffused simultaneously to a junction depth of about 1 μm. Contact windows are now opened and a 0·8 μm thick aluminium layer is evaporated from an electron gun. The sixth and final masking step defines the interconnection pattern. Note that the number of masks used, 6, is equal to that of the conventional buried layer bipolar process. The static divider structure shown in figure 4 is free of any essential logical hazards and it works independently of the delay associated with each gate. Furthermore, the inverted input variable is not required which avoids the use

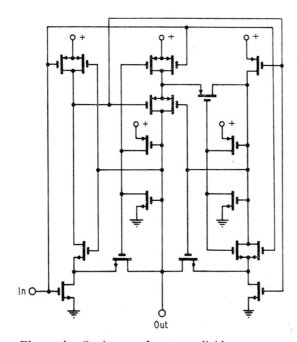

Figure 4. Static CMOS frequency divider stage.

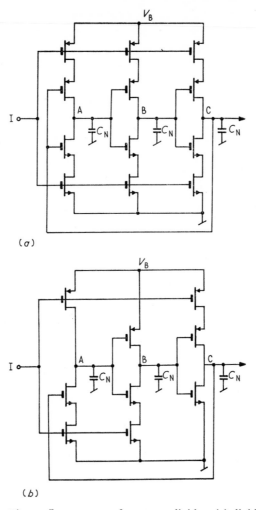

Figure 5. CODYMOS frequency divider: (*a*) divider by 3; (*b*) divider by 2.

of an inverter to drive the first stage and thus contributes to a reduction of the power drain.

Complementary dynamic metal–oxide–semiconductor (CODYMOS) frequency dividers are obtained by interconnecting p- and n-channel MOS transistors according to their logical equations. Figure 5 shows the diagram of a divider by 3 (Oguey and Vittoz 1973). It consists of 3 stages, A, B and C of 2 p-channel and 2 n-channel transistors each. The input signal drives 2 complementary transistors of each stage, whereas each internal variable drives two complementary transistors of the following stage. This structure is ripple and hazard free as every change of the input variable I results in a corresponding change in one output variable only.   We therefore expect the structure to be very fast.   The highest frequency of the input variable is approximately given by

$$f_{max} \simeq \frac{\beta_p(V_B - V_{Tp})^2}{8\,C_N V_B}$$

where β_p refers to the conduction factor and V_{Tp} to the threshold voltage of the p-channel transistor. V_B is the supply voltage and C_N the highest node capacitance.   The current consumption of a simple divider stage is given by

$$I = f(C_N + C_{IN})V_B.$$

Assuming that $\beta \sim W/L$ and $C_N \sim WL$,† with W and L the channel width and

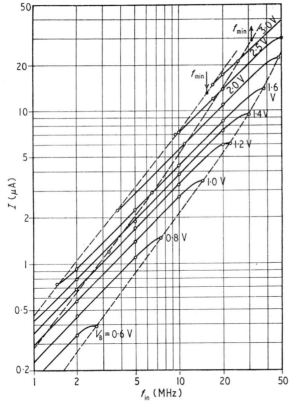

Figure 6.   Current drain against frequency, and frequency limits of an integrated CODYMOS frequency divider.

† Actually C_N should approach an asymptotic finite value as L tends to zero.   The simplified relation neglects a variable fraction of the drain junction capacitance as well as the interconnection capacitance.

length respectively, we conclude that at a given frequency the current is proportional to WL, whereas the frequency limit is proportional to L^{-2} and roughly independent of W. Both current drain and frequency limit may be improved by a reduction of the appropriate dimensions. This circuit has been integrated using the silicon gate technology we have described earlier. The dimensions of p- and n-channel transistors are identical (nominal mask dimensions of $W=8$ μm and $L=5$ μm). Typical measured values are 35 μA V^{-2} for β_p and -0.3 V for the p-channel threshold voltage. At a supply voltage of 1·2 V, the series connection of two transistors is capable of drawing a current of 7·1 μA. A node capacitance C_N equal to 0·12 pF is charged in 20·3 ns, corresponding to an input frequency f_{max} equal to 24·6 MHz. The input capacitance C_{IN} is equal to 0·18 pF, which includes the gate capacitance of the driver stage. The current drain should amount to 0·42 μA MHz^{-1} at a supply voltage of 1·4 V. In figure 6 we have plotted for a high performance unit the current drain of one stage, including an input driver as a function of the input signal frequency.

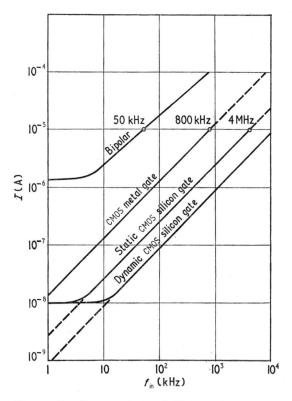

Figure 7. Current drain of 15-stage frequency divider chains implemented by means of different circuit concepts and technologies ($V_B=1\cdot35$ volts).

This circuit attains a maximum frequency in excess of 20 MHz at a supply voltage of 1·2 V and draws less than 4 μA at 10 MHz and 1·4 V. Near the maximum frequency the current levels off owing to a decrease in internal signal amplitudes, and evidently, in order to achieve reliable operation, this region is to be excluded. The dynamic nature of this circuit excludes its use at low frequencies. At an intermediate frequency, its behaviour may depend on initial conditions, but within a wide range of frequencies and supply voltages the circuit divides correctly for any initial condition.

In the lower part of figure 5 a divider by 2 is shown. It was originally derived from its static counterpart we have shown earlier, but we may also look at it as a contraction of the divider by 3. The elimination of 3 transistors transforms 2 of the 6 stable states into transitory states, so that the dividing ratio becomes 2. The performance of this circuit is only slightly inferior compared to the divider by 3. On the other hand, it is entirely insensitive to initial conditions and divides correctly down to frequencies below 1 kHz. To summarize this section we have plotted in figure 7 the current drain of 15-stage frequency

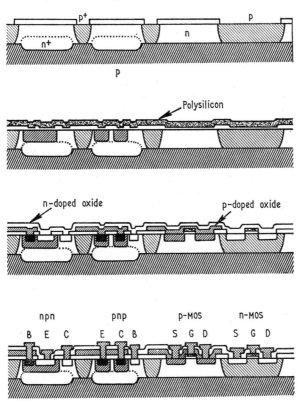

Figure 8. Outline of fabrication sequence for the simultaneous integration of bipolar and MOS transistors.

dividers as a function of input frequency at 1·35 V, implemented with different circuit conceptions and technologies. The particular frequencies indicated on the graph correspond to a current drain limit of 10 μA, which in view of applications to watches may be considered an optimistic upper limit. As we have noted earlier, bipolar techniques are useful up to 32 kHz. The metal gate CMOS referred to here does not make use of ion implantation in order to achieve selfaligning. Static silicon gate CMOS circuits as shown here may be used up to 4 MHz. (The same level of performance might be expected from any of the ion-implanted selfaligned structures.) As one would expect, the winner is a dynamic silicon gate CMOS circuit. State-of-the-art technology allows 10 MHz to be reached. Further improvements in all cases could be realized by using the well known PLANOX technology. Still higher levels of performance are probably the preserve of the semiconductor on insulating substrate technologies.

In electroluminescent diode display systems, the output transistors have to supply relatively high peak currents and a MOS solution might demand an inordinate fraction of chip real estate. While in many cases a hybrid solution might be optimal, one might conceive situations where the simultaneous integration of MOS decoding circuitry and bipolar driving transistors could present the most advantageous solution. Using our standard CMOS process as a starting point, we set out to find a technology that would enable us to integrate simultaneously npn and pnp bipolar transistors as well as n- and p-channel MOS transistors. This work will be described elsewhere (Darwish and Taubenest 1974) and we limit ourselves to a brief description of the basic steps of the fabrication process. The buried n^+ layers in figure 8 are diffused in a p-type (100) substrate. Their function is to provide a low saturation voltage for the vertical npn transistors and at the same time to increase the emitter efficiency of the lateral pnp by reducing vertical hole injection. By thermal decomposition of silane, an n-type epitaxial layer is grown to a thickness of approximately 7 μm. Its resistivity is in the range of 2 ± 0.5 ohm cm.

Table 1. Characteristic properties of the MOS transistors

Type of transistor	$V_{\text{th. extrapol.}}$	β	$\beta W/L$
n-channel	0·68 V	13	13
p-channel	−0·80 V	3·8	3·8
Thick oxide	5·45 V		

(W = channel width; L = channel length; W/L = 1 for the test transistors)

Table 2. Characteristic properties of the bipolar transistors

Type of transistor	h_{fe} at $I_B = 1$ μA and $V_{CE} = 1$ V	BV_{EBO} (V)	BV_{ECO} (V)	BV_{BCO} (V)
npn	150	13	10	38·5
pnp	8	38·5	35	38·5

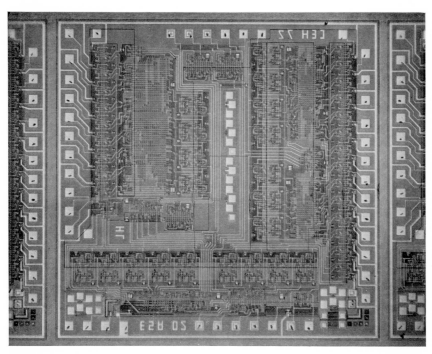

Figure 9. Photomicrograph of a 1200 CMOS transistor circuit providing frequency dividing, decoding and driving functions for a digital electrochromic display.

(*facing page 40*)

A 1 μm thick oxide is now grown and windows are opened for the p-isolations and p-wells. These p-type regions are diffused in a sealed capsule for 10 hours at 1200 °C. The p-type source wafers have a resistivity of 0·3 to 0·4 ohm cm.

The base diffusion of the vertical npn's is accomplished by means of a boron-doped oxide diffusion step. Junction depth is 2·6 μm and the sheet resistivity around 600 $\Omega/\square$. Windows defining the source–gate–drain regions of both the p-channel and n-channel MOST are opened simultaneously with the n^+ and p^+ contact regions of the bipolar elements.

From this point on, the fabrication sequence is identical to that of the silicon gate CMOS process described earlier. Tables 1 and 2 summarize some of the parameters.

The current gain of the lateral pnp could be improved by reducing the base width, as is indicated by the relatively high value for BV_{ECO}, and by replacing antimony by arsenic for the buried layer in order to increase the diffusion length for the minority carriers. Another reason for the rather low h_{fe} values obtained is related to the asymmetrical bar structure of these lateral transistors.

Figure 9 is a photomicrograph of a rather complex silicon gate CMOS circuit that provides frequency dividing, decoding and driving functions for a digital electrochromic display. The segment-driving output MOS transistors have to supply a given current within $\pm 10\%$.

3. Electronic watch display techniques

In table 3 we have listed some of the existing display techniques.

Table 3. Display techniques.

Active Displays	Passive Displays
LED Digital	Liquid Crystals
LED Analog	Electrochromism
Incandescent Lamp	Ferroelectrics
Bulk Electroluminescence	Electrophoretics
Cathodoluminescence	
Plasma	

Up to now only one kind of active and one type of passive display has been commercially exploited in electronic wristwatches: digital electroluminescent diode displays and various forms of liquid crystal displays. As far as active displays are concerned, it would seem that the electroluminescent diode is the only serious contender. As for other passive displays, research is being pursued in the field of electrochromism, ferroelectrics and electrophoretics (Meitzler and Kurtz 1973). An electrochromic display based on oxidation–reduction reactions of an organic compound of the viologen family has been presented recently by a group at Philips Research Laboratories (Schoot *et al* 1973). Contrast ratios of 20 to 1 are achievable and because one deals with light absorption phenomena,

D

contrast is independent of the angle of viewing. Writing time decreases with increasing voltage and is in the 10 to 100 ms range for good contrast. The energy necessary for one write–erase cycle is about 4 mJ cm^{-2}. Owing to the memory property of the device, displaying the minute would require a power consumption of 70 μW cm^{-2}. Life data is very limited at present and predictions concerning the usefulness of this display in practical applications are somewhat premature.

Electro-optic effects in ferroelectric ceramics, like hot-pressed lead-zirconate-titanite ceramics [Pb$_{0.99}$Bi$_{0.02}$(Zr$_{0.65}$Ti$_{0.35}$)$_{0.98}$O$_3$] have been known for a number of years. The high voltages necessary to electrically control the birefringence in a ferroelectric ceramic should probably rule out their applications in wrist watch displays.

Electrophoretic displays are reflective devices using as their basic mechanism the electrophoresis of charged pigment particles in a suspension. Reflective contrast ratios greater than 30 to 1 have been demonstrated with 100 V DC at the cell. Power dissipation is about 35 μW for displaying a 7×3 mm numeral '8'. This figure is about two orders of magnitude above that for a liquid crystal display.

Turning to the liquid crystal display we compare in table 4 a number of properties for the dynamic scattering, distortion of aligned phases and twisted nematic modes.

Table 4. Comparison of liquid crystal display modes

	DS	DAP	TN
Power	highest	low	low
Operating voltage	highest ($\geqslant 15$ V)	low ($\geqslant 4$ V)	lowest ($\geqslant 1$ V)
Expected cell life	lowest	high	high
Washout (undesirable reflections)	yes	no	no
Angular viewing range	lowest	higher	highest
Contrast	good	better	best
Response time (typical values)	300 ms	200 ms	200 ms
Cost	—	—	—

Taking everything into consideration, the twisted nematic mode comes out as a winner.

In the last line concerning cost we have deliberately left blanks. It might be tempting to consider the dynamic scattering mode display as lowest in cost owing to the absence of polarizers. In fact, this advantage is probably more than offset by the necessity to incorporate a DC to DC converter. Cell life in general is still a real problem. The longest cell life guaranteed by one manufacturer is 3 years for a twisted nematic display.

R A Soref at the Sperry Research Center has recently come up with a rather interesting interdigital field effect liquid crystal display (Soref 1973). He

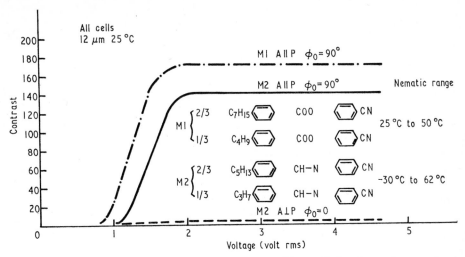

Figure 10. Variation of contrast with applied voltage for a twisted nematic mode field effect liquid crystal display (Boller *et al* 1972).

reported also on a new reflective viewing mode, which uses one circular polarizer and a diffuse metallic reflector. Under ambient illumination, addressed areas appear white on a dark background. The fact that the two sets of electrodes are confined to one single plane may offer an additional degree of freedom in the conception of integrated decoding–display units.

It is obvious that a watch liquid crystal display should preferably work at a voltage around 1·5 V. The elimination of the DC to DC converter, necessary in all dynamic scattering mode watch displays, results in an appreciable reduction of space, power consumption and cost. Twisted nematic mode cells have been successfully operated at conveniently low voltages (Boller *et al* 1972). Figure 10 shows the variation of contrast with applied voltage for two nematics mixtures. The threshold of mixture M2 is somewhat higher but in view of the considerably larger temperature range it would seem to offer an excellent overall solution.

The number of interconnections necessary between the decoding–driving circuit and the display, which is particularly large in static displays, could be vastly reduced by integrating the two functions into one monolithic block. Robert and Borel (1971) have successfully built an integrated display using a conventional MOS circuit and a layer of liquid crystal operated in the dynamic scattering mode placed right on top of the circuit (figure 11). Each character of their display consists of a 5 by 7 dot matrix. All data corresponding to the display of a given numeral are entered serially into the shift register and stored. One bit is implemented by a conventional two clocks, three inverter shift register driving one channel of a 35 outputs multiplexer.

Brody and his co-workers at Westinghouse (Brody *et al* 1973) recently built

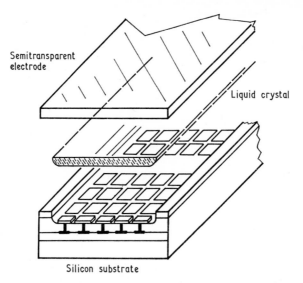

Figure 11.   Integrated MOS decoder/liquid crystal display block (Robert and Borel 1971).

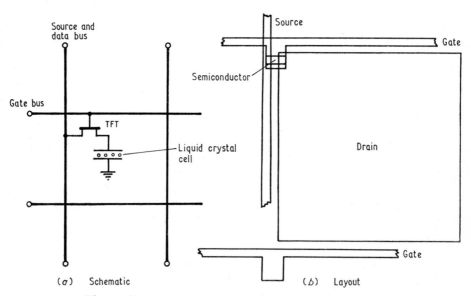

Figure 12.   Hybrid thin film transistor decoder/liquid crystal display.

and operated an integrated 14 000 picture element flat screen panel 36 square inches in area by a combination of thin film transistor and nematic liquid crystal technology.

In figure 12 we have shown a simple matrix design which is designed to operate as follows. Amplitude information is stored in a row of peripheral capacitors and during flyback the capacitors are discharged into all columns simultaneously, while the transistors of one row are turned on by a vertical scanner. The stored charges will then divide between the peripheral capacitor and the picture element capacitor, thus producing voltages across the liquid crystal in proportion to the stored amplitudes. The display which is able to reproduce a gray scale was originally devised to display a TV picture, but the basic idea might provide one economical solution for watch and clock displays. The entire thin film matrix can be deposited through metal aperture masks in a single pump-down cycle. For the twisted nematic mode operation the required homogeneous surface alignment over the large areas was obtained by evaporating approximately 70 Å of silicon monoxide or gold at a glancing angle. This technique evidently is highly compatible with the rest of the processing and does not involve handling of the finished circuit.

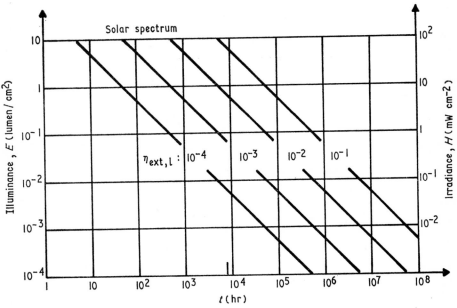

Figure 13. Lifetime, in hours, of an electrochemical energy source of 100 mWh activating a $4 \times 10^{-4} \text{cm}^2$ LED under various irradiation levels. The lifetime is given by $t = 3 \times 673\epsilon\,\eta_{\text{ext},1}/EA_{\text{d}}$. LED parameters: emitting surface $A_{\text{d}} = 4 \times 10^{-4}\text{cm}^2$; luminous external efficiency, $\eta_{\text{ext},1} = \eta_{\text{ext}}R(\nu)$ where η_{ext} is the external efficiency and $R(\nu)$ is the eye response.

It is probably a safe guess to say that the simultaneous integration of decoding and display functions of which we just presented two examples will be a key element in all future development of electronic watches.

All commercially available watches using an electroluminescent diode display are of the digital variety.   In some cases, the diode brightness is adjusted automatically as a function of the ambient light level, corresponding to extreme values of 3 mA and 100 mA of current.   In order to assure a halfway acceptable life for the battery, the information is displayed on demand only.   Depending on the user's habits as well as the average ambient light level, large variations of battery life will unavoidably occur.   Let us have a look at the problem of a continuous display powered by a chemical battery.   From figure 13 we conclude that in the low illumination range, say below 10^{-3} lumen/cm^2 an electrochemical energy source is able to power a 10^{-4} efficiency commercial LED continuously for many years.   At 10^{-1} lumen/cm^2 a LED with an efficiency of 10^{-3} can be powered by a battery for about one year.   A continuous display is conceivable if one assumes that these ambient light levels occur only infrequently and if one chooses an analog type of display, using only a very limited number of active diodes.   In bright open sunlight, corresponding to 10 lumens/cm^2, a chemical battery can drive a state-of-the-art diode for about 20 hours.   This in turn amounts to only 40 minutes for a digital display.

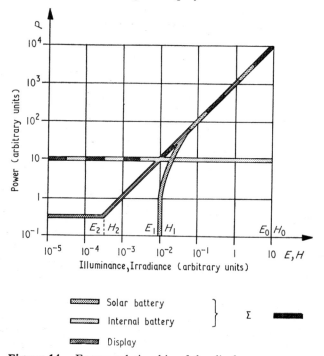

Figure 14.   Energy relationship of the display.

The system presented here (Vuilleumier 1971) incorporates a continuous analog display whose power is obtained from the ambient irradiation by means of a solar cell battery. An internal electrochemical energy source is used at light levels below a certain threshold and the system automatically adjusts the LED brightness to give a display characterized by a constant contrast.

The energy relationship between the power sources is shown in figure 14. The upper area of the figure corresponds to high light levels, where the solar cells act as the main energy source. When the ambient illumination approaches E_1, the solar battery ceases to produce power and its open circuit voltage approaches the voltage required to activate the LEDs. The system utilizes now the electrochemical energy source to maintain a constant contrast display. At E_2 the system goes to a constant power mode which is useful at zero ambient light level.

The number of diodes that can be powered by a solar battery with a surface of 1 cm^2 is shown in table 5. Other relevant parameters are the solar battery efficiency and the external luminous efficiency of the light-emitting diodes.

Table 5. Number of LEDs activated by a solar cell, where N_d is the number of LEDs and η_s is the efficiency of the solar battery

η_s	10^{-1}	5×10^{-2}
N_d ($\eta_{ext,1} = 10^{-1}$)	500	250
N_d ($\eta_{ext,1} = 10^{-2}$)	50	25
N_d ($\eta_{ext,1} = 10^{-3}$)	5	2
N_d ($\eta_{ext,1} = 4 \times 10^{-4}$)	2	1

Figure 15 shows the block diagram of the ambient powered part of the system and illustrates the electronic functions. The dynamic range covered is about 2000 and the system takes into account the fact that the diode efficiency decreases at high current densities owing to thermal effects and that it also decreases at low current densities owing to the preponderance of nonradiative recombination currents. A capacitor is charged by a number of solar cells. These cells produce a voltage not less than V_R when irradiated by a light level equal to E_1 where V_R is so chosen as to provide an efficient diode excitation. The moment the voltage across C exceeds V_R, the comparator applies a signal L to the logic circuit, indicating that an efficient discharge of C can take place. The logic circuit also receives a clock pulse T, which sets the pulse repetition rate of the display. Let us keep in mind that when the pulse repetition frequency is greater than the reciprocal persistence of the eye the display will appear to be continuous. The logic circuit generates a signal F which acts on the selected diode switches through a series of AND gates, whose outputs F_i control the power switches. The signal F holds the power circuit closed until the voltage across C drops below V_R. The circuit functions as a voltage regulator for the solar cells and associated capacitor, and generates a pulsewidth modulated current supplied

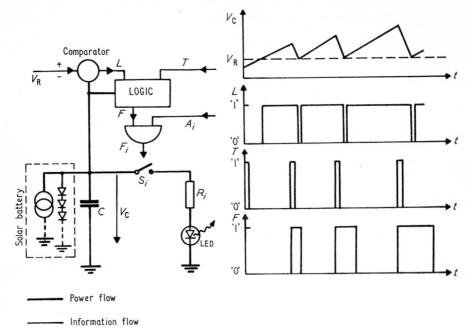

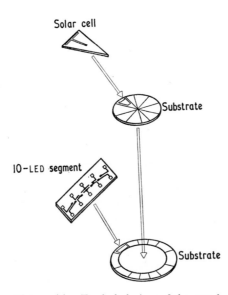

Figure 15.   Block diagram of the ambient powered section of the system.

Figure 16.   Exploded view of the watch face assembly.

to the diodes. The pulse duration is proportional to the charge accumulated in C and the energy supplied by the capacitor is proportional to the ambient light level and therefore to the light emitted by the LEDs.

An exploded view of the watch face assembly is shown in figure 16. Anodized aluminium was chosen as a substrate material. Mounted on the periphery, we have a dozen segments consisting of 10 LEDs each. The central portion is occupied by the 12-cell solar battery. The solar cells fabricated in planar technology were specifically made for the reasons summarized in table 6.

In view of the rather small currents, the relatively large series resistance is of little importance, whereas the twenty-fold improvement in parallel resistance is extremely important for low level applications. The lower the solar cell leakage

Table 6. Solar cell design parameters

Parameter	Watch systems	Commercial cells (space applications)
Conversion efficiency	Adapted to domestic illumination	Adapted to solar spectrum beyond earth atmosphere
$\left.\begin{array}{l} R_p \\ R_s \end{array}\right\}$(cell surface $0\cdot12$ cm²)	High ($\geqslant 2$ MΩ) Low (<1 Ω)	Lower ($0\cdot1$ MΩ) Very low ($<0\cdot1$ Ω)
Geometry	Specially adapted to watch face geometry	Big rectangles
Radiation hardness	Not critical	Critical

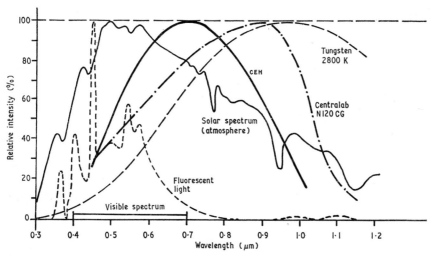

Figure 17. Emission spectra of three typical light sources; solar spectrum, fluorescent light and tungsten light at 2800K, and relative spectral response of two types of solar cells.

current, the lower will be the current level at which the chemical energy source is needed.   Figure 17 shows the emission spectra of three typical light sources: solar spectrum, fluorescent light and tungsten light at 2800K, together with the relative spectral response of two solar cells.   Compared to a commercial cell, our n^+-p cell with a junction depth of 0·5 micrometres shows a desirable shift towards shorter wavelengths, thus approaching the spectral response of the human eye.

4. Conclusion

State-of-the-art micropower integrated circuits, in particular CMOS circuits, provide a considerable degree of freedom to the designer of electronic watches. With power consumption being at a low 0·4 μA MHz^{-1} level, the incorporation of inexpensive, low temperature coefficient and shock resistant AT-quartzes in the time base has become feasible.   It should be stressed that these performance levels are the result of considerable progress in the areas of circuit principles as well as technology.   Micropower circuits have also made possible the implementation of logic circuit functions beyond the classical function of frequency division, for example decoding for electronic displays or memory functions. These developments may well point the way to multifunctional watches.   There is also a trend to simplify certain fabrication steps, like the accurate adjustment of a quartz at the cost of a minor increase in the complexity of the electronic system.   A vast amount of work remains to be done in the field of electronic displays and we believe that the integration of decoding and display functions in hybrid or monolithic form will play a key role in future developments of electronic watches.

References

Boller A, Scherrer H, Schadt M and Wild P 1972 *Proc. IEEE* **60** 1002
Brody T P, Asars J A, Dixon G D, Escott W S and Hester W A 1973 *Westinghouse Research Labs. Report* 73-9 Fl-Liquid-Pl
Darwish M and Taubenest R 1974 *J. Electrochem. Soc.* to be published
Forrer M P 1972 *Proc. IEEE* **60** 1047
Meitzler A H and Kurtz S K (ed) 1973 *Proc. IEEE* **61** 804
Oguey H and Vittoz E 1973 *Electron. Lett.* **9** 386
Robert J and Borel J 1971 *Technical Note* LETI/ME No 728
Schoot C J, Ponjee J J, van Dam H T, van Dorn R A and Bolwijn P T 1973 *Appl. Phys. Lett.* **23** 64
Soref R A 1973 *Society for Information Display Symposium, New York, May 1973* paper 4.1
Thommen W and Ruegg H 1971 *Eurocon, Lausanne, October 1971* paper F2-5
Vittoz E, Gerber B and Leuenberger F 1972 *IEEE J. Solid St. Circuits* **7** 100
Vittoz E, Hammer W, Kiener M and Chauvy D 1971 *Eurocon, Lausanne, October 1971* paper F2-6
Vuillemier R 1971 *Eurocon, Lausanne, October 1971* paper F4-5

Layered dielectrics in the MOS technology

P Balk

Institut für Halbleitertechnik,
Technical University, 51 Aachen, German Federal Republic

Abstract

This paper presents a review of the characteristics of SiO_2–PSG, SiO_2–Si_3N_4 and SiO_2–Al_2O_3 dielectrics in MOS systems and their use in the FET technology. After a brief outline of the fundamentals on conduction in layered insulators and band diagrams for such systems, the pertinent properties of the component materials and that of the Si–SiO_2 interface are discussed. The layered gate exhibits a fundamental instability; it is shown which parameters contribute to this problem and how to arrive at combinations with optimal stability. It is also shown that there are basic limitations to the use of such structures for very thin gate capacitors. A discussion of the literature on MNOS and MAOS non-volatile memory devices leads to the conclusion that our insight into the detailed mechanism of operation is still limited, and that not sufficient pertinent data are available. Most likely both structures exhibit a similar shift of their logical states ('window collapse') under continued pulsed operation which makes RAM application difficult. Some new possible approaches are indicated.

1. Introduction

Control of the properties of the gate area is an essential requirement for the successful practice of FET technology. As the result of over a decade of research and development, metal–SiO_2–silicon (MOS) gate structures with well-defined characteristics can routinely be fabricated. Annealing conditions leading to the elimination of fixed oxide charge and interface states at the Si–SiO_2 boundary (Revesz and Zaininger 1968, Montillo and Balk 1971) essential for obtaining reproducible threshold (V_T) values and high transconductances, have been established. It is also known that Na impurities will lead to V_T drift under positive bias (Snow *et al* 1965), whereas even in pure SiO_2 films a so-called slow trapping instability may occur, particularly apparent under negative bias and resulting in an increase in interface states and surface charge (Miura and Matukura 1966, Goetzberger and Nigh 1966, Deal *et al* 1967, Hofstein 1967). The clean processing facilities, required for avoiding contaminants in the device fabrication process, have also yielded better gate oxide integrity and have greatly reduced the incidence of low voltage dielectric breakdown.

The unique and favourable properties of the Si–SiO_2 system notwithstanding, a search for alternative insulators had already begun in the mid-sixties. Particularly a better protection of the devices against Na contamination than SiO_2 offered was desired. Since other insulators did not appear to give the favourable interface characteristics that SiO_2 does at the silicon contact, and because diffusion of a component of the film into the silicon during high temperature processing would affect the carrier density near the silicon surface, a combination of SiO_2 with a second dielectric layer seemed called for. The results of this effort were the SiO_2–PSG (phosphosilicate glass) (Kerr *et al* 1964), the SiO_2–Si_3N_4 (Tombs *et al* 1966) and the SiO_2–Al_2O_3 systems (Nigh *et al* 1967). This last system has the additional attraction that it reduces the threshold of a p-channel FET device and makes an enhancement mode n-channel device possible.

Whereas these new materials were first introduced with much enthusiasm, it was soon discovered that they exhibited their own specific contributions to threshold instability. Considerable effort was made to understand and minimize these problems. Also, it was subsequently realized that these effects may be utilized to advantage for non-volatile information storage in the metal–Si_3N_4–SiO_2–Si (MNOS) (Pao and O'Connell 1968, Sugano *et al* 1968) and metal–Al_2O_3–SiO_2–Si (MAOS) (Nakamura *et al* 1970, Balk and Stephany 1970) structures.

In the continued quest for increased density and performance in FET devices and circuits (Broers and Dennard 1973) the thickness of the gate insulator is a very important factor. It seemed quite possible that the fabrication yield and reliability problems, most likely resulting from a reduction of the insulator thickness from the value of 1000 Å which is probably current for presently manufactured devices, might be alleviated by using layered structures. Thus the last several years have shown a continued research effort in the area of gate dielectrics.

This paper will present a review of our present level of understanding in this field. No effort towards presenting an exhaustive bibliography will be made. Since a recent review by Kahng and Nicollian (1972) has specifically dealt with the physics of multilayer FET gates, the present paper will focus particularly on materials properties and the technological aspects of the problem.

After a brief discussion of the phenomena playing a role in the behaviour of layered dielectric gates, the properties of the major insulators considered for MOS application (SiO_2, PSG, Si_3N_4 and Al_2O_3) and those of the Si–SiO_2 interface will be reviewed. The next section deals with the observed characteristics of double insulator layers. MNOS and MAOS memory structures, a subject of considerable publication activity in the past four years, will be discussed separately. The concluding section will attempt to summarize the present state of affairs, and to give an outlook for the further development of this sector of MOS technology.

Not discussed in this paper are the effects of particle and x ray radiation on insulators. These phenomena, which are not only important for space or military applications but also play a role in modern semiconductor processing techniques (ion implantation, sputtering, electron beam methods), had to be left out to limit the size of the review.

2. Electrical phenomena in layered dielectrics

2.1. Conduction in layered structures

When applying a voltage V_a to a layered dielectric structure (figure 1; insulators I_1, I_2, thicknesses x_1, x_2, dielectric constants ϵ_1, ϵ_2) currents J_1 and J_2, whose magnitudes are determined by the electric fields E_1 and E_2 in I_1 and I_2 respectively, will flow through the insulators (Sewell *et al* 1969). These fields are related to the applied voltage by

$$E_1 x_1 + E_2 x_2 = V_a. \tag{1}$$

The boundary condition is the principle of continuity of dielectric flux, which gives for a charge-free system the expression

$$E_1 \epsilon_1 - E_2 \epsilon_2 = 0. \tag{2}$$

Currents J_1 and J_2 are generally unequal. This leads to a build-up of charge Q_i in the system near the insulator–insulator interface

$$\frac{dQ_i}{dt} = J_2(E) - J_1(E), \tag{3}$$

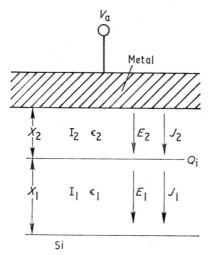

Figure 1.   Schematic diagram of a MI₂I₁S structure.

and to a change of the fields according to the requirement of continuity of flux

$$E_1\epsilon_1 - E_2\epsilon_2 = Q_i, \tag{4}$$

until a steady state with $J_1 = J_2$ will be reached.   Here it has been assumed that the density of trapping states near the I_1I_2 interface is sufficiently large to accommodate the charge Q_i.   The flatband voltage V_{FB} of an MI₂I₁S capacitor and V_T for a FET device will show a shift (compared to the charge-free case) proportional to the magnitude of Q_i:

$$\Delta V_{FB} = -Q_i \frac{x_2}{\epsilon_2}. \tag{5}$$

2.2. Possible conduction mechanisms

Electronic conduction through insulators exhibits a very pronounced field dependence.   The current at room temperature is generally an exponential function of some power (1/2 or 1 in the cases of interest in this paper) of E (Simmons 1970).   These conduction models are illustrated in figure 2.   A common mechanism is the field-assisted thermal excitation of carriers from traps into the conduction band (Poole–Frenkel effect).   However, in some cases the current is not determined by the bulk resistance of the insulator, but by the rate at which carriers are supplied from the metal or semiconductor electrode, or from the neighbouring insulator.   Fowler–Nordheim tunnelling into the valence band of the insulator, direct tunnelling into traps, or for very thin films (typically less than 50 Å thick) tunnelling through one insulator into the next are the mechanisms that have been observed.   A problem in the experimental determ-

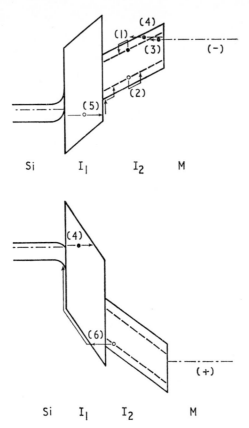

Figure 2. Current transport in insulators by electrons and holes: (1), (2) Poole–Frenkel effect; (3) tunnelling into traps; (4) Fowler–Nordheim tunnelling; (5) modified Fowler–Nordheim tunnelling; (6) tunnelling from traps.

ination of the field dependence of the insulator current is that trapped carriers, injected during the measurement, will affect the field distribution in the system. This is in agreement with the observation that the current at a fixed voltage in a material with traps will vary with time.   An analysis of this problem has been given by Walden (1972).   Thus, the use of the experimentally determined steady state field and temperature dependencies of the currents can only be used in the calculation of the steady state value of Q_i.   Utilizing these dependencies for calculation of the time dependence of Q_i may lead to considerable error.

2.3. Band diagrams, trapping states and barrier energies

Interpretation of the experimentally observed conduction behaviour in a MI_2I_1S system requires knowledge of its band structure.   It appears that the energy band concept does not only apply to single crystals, but also to amorphous and polycrystalline insulating films (Lamb 1967). Trapping states are very common in these wide band gap materials. They are often related to non-stoichiometry or dopants, but may also occur intrinsically in pure materials of stoichiometric composition (Revesz 1973). Band structures, as the one drawn schematically in figure 2, have not been determined directly for the systems to be discussed in this review, but have been constructed from barrier energies (obtained from internal photoemission), band gap and vacuum work function data. So far insulator–insulator barriers have not been determined directly. The estimation of the metal–insulator barrier for one metal from the measured barrier height with a second metal and the difference in work functions for both metals appears questionable. It is unlikely that the surface dipoles of both metals would change by the same amount upon being brought into contact with the insulator. Also, the shape and thus the effective height of a metal–insulator barrier may be affected by the formation of an ionic space-charge layer due to interfacial reaction and diffusion of metal ions into the insulator. This seems to be borne out by the observation that the flatband voltages in MOS systems, in which oxide charge centres and interface states had been eliminated before metallization, are nonetheless affected by low temperature annealing (P Balk 1970 unpublished). In fact, it appears that the relative barrier energies for different metals obtained from C–V measurements, show a better correlation with the electronegativities of the metals than with their vacuum work functions (Deal *et al* 1966). Poor correlation with the relative values of the metal work function was also noted by Kalter *et al* (1971) in C–V measurements with various electrodes on different insulators. Charged impurities will, like any charge near the barrier, affect the barrier energies (DiStefano 1971, Brews 1973). For these reasons the band diagrams for layered structures shown in a later section have to be treated with some reservation.

3. Insulator properties

3.1. Silicon dioxide

Since SiO_2 is a component of all layered dielectric systems to be discussed, a review of the pertinent properties of these films appears to be in order. As used in the gate of FET devices amorphous SiO_2 films are generally prepared by thermal oxidation in oxygen of the silicon substrate (Deal and Grove 1965, Revesz and Evans 1969). The reproducible growth of very thin films, as required in MNOS-type non-volatile memory devices, requires particular care. They may be obtained by oxidation in O_2, steam or in NO with N_2 or H_2 (Goodman and Breece 1970, Aboaf 1971, van der Meulen 1972, Ghez and van der Meulen 1972). The attainable reproducibility in the 20–30 Å range appears to be ± 1 Å. Very thin SiO_2 films can also be grown at low temperatures in HNO_3 (Oakley and Godber 1972). The material has a bandgap of

8·9 eV (DiStefano and Eastman 1971a) and the relatively small dielectric constant of 3·82 (Sprague *et al* 1962). Even when stoichiometric in composition, SiO_2 may exhibit hole trapping due to the occurrence of non-bonding p-valence band levels, originating in the O^{2-} ion (DiStefano and Eastman 1971b), or of π bands, resulting from the overlap of 3d Si orbitals with 2p O orbitals (Revesz 1973). These levels, which would be deep in the SiO_2 gap or would even form a part of the valence band system, could be responsible for the pronounced tendency of SiO_2 films on silicon to become positively charged during oxidation.

Other oxide levels are probably related to lattice imperfections and foreign atoms. Mitchell and Denure (1973) believe that the three centres observed in their cathodoluminescence studies are due to trivalent Si, to Na and to H. Electron trapping related to the presence of water in SiO_2 has been reported by Nicollian *et al* (1971). The energies of these centres are not known. A distribution of states, centred around 2·1 eV below the conduction band edge was reported by Thomas and Feigl (1972) from measurement of the optical release of trapped electrons. The origin of these states is not clear.

Probably because of its relatively high level of perfection, SiO_2 forms a very poor barrier against the motion of positively charged contaminants in electric fields. Many studies have been devoted to Na^+ drift. Studies on sputtered SiO_2 films suggest that the Na^+ motion is determined by the rate of emission from oxide traps, and that it will move more slowly in films with a high density of imperfections (Hickmott 1973). The transport of mobile ions (presumably Na^+) from the metal–SiO_2 to the silicon–SiO_2 interface was found to be trap emission controlled, the traps being located near or at the metal interface. The fraction of ions that can be mobilized is temperature dependent. This finding points to a distribution of trap depths. Transport in the opposite direction is more rapid and bulk controlled (Kuhn and Silversmith 1971). It has also been reported that the H^+ ion will drift in electric fields (Hofstein 1967), but this contention has recently been disputed (Raider and Flitsch 1971).

Na may be incorporated into SiO_2 films as an ionic and as a neutral species (Kuper 1969). Presumably only the ionic form will contribute to threshold shifts in FET devices. However, it is not quite clear under what conditions transformation of one form into the other may take place. Thus great efforts have been made to avoid Na contamination during the preparation of the MOS gate. A powerful method to prepare Na-free oxide films is to grow these layers in O_2 with a few per cent of HCl added to the gas stream. Not only does the HCl improve the purity of the films, but particularly at high oxidation temperatures (at least 1100 °C) Cl will be incorporated into the oxide near the Si–SiO_2 interface. These built-in atoms will passivate the surface by trapping Na^+ ions; the positive charge of the Na^+ is being transferred to the silicon (Kriegler 1973). Subsequent high temperature annealing in an inert gas will gradually release the Cl, and thus reduce the passivation effect.

E

One of the attractive properties of thermal SiO_2 films is their high intrinsic breakdown strength of close to 10^7 V cm^{-1} (Fritzsche 1967, Chou and Eldridge 1970). Thus, it is essential to avoid processing procedures in substrate cleaning and oxidation that lead to defect-related low-field breakdown. The technological variables affecting the breakdown behaviour have been investigated in detail by Osburn and Ormond (1972). The conduction in SiO_2 films is limited by the injection of electrons via Fowler–Nordheim tunnelling at the electrodes (Lenzlinger and Snow 1969, Osburn and Weitzman 1972). At high fields most likely impact ionization takes place, leading to breakdown via an electronic mechanism. The drifting of mobile Na$^+$ ions in the film will not only effectively lower the barrier for injection, which will permit higher current flow and may lead to breakdown (DiStefano 1973), but it will also enhance the internal field in the bulk of the oxide until it reaches the critical value for breakdown (Osburn and Raider 1973).

3.2. *The Si–SiO$_2$ interface*

Because of its high practical interest the electrical characteristics of the Si–SiO$_2$ interface have been extensively investigated. One of the most remarkable properties of this interface is that electrically active centres in the SiO_2 or at the Si–SiO$_2$ boundary can be reduced by proper annealing treatments to below 10^{10} cm^{-2}, which amounts to one centre in 10^5 interface atoms. There is no general agreement in the question what structures correlate with oxide charge centres and fast interface states. However, it has been established that both centres are located within 20 Å from the Si–SiO$_2$ interface (Powell and Berglund 1971). There are basically two schools of thought regarding interface states: Goetzberger *et al* (1966) have proposed that interfacial traps could be induced by point charges in the oxide near the interface and Hickmott (1972) has stressed the similarity in annealing behaviour of these states and a band observed in RF sputtered quartz. Barruel and Pfister (1972) have observed a correlation between the amount of surface charge and the density of electron traps near the centre of the SiO_2 bandgap, as obtained from photoconductivity measurements. They feel that these centres are O interstitials or Si vacancies. Other workers have suggested that the states are trivalent silicon atoms located directly at the interface (Kooi 1965, 1966, Montillo and Balk 1971). It is essential for obtaining high transconductance values to remove these centres. This is done most conveniently by annealing in H_2 or water vapour at moderate temperatures (200–500 °C). Most likely Si–H bonds are formed under these conditions.

Oxide charge introduced in the thermal oxidation process may be removed by high temperature (greater than 1000 °C) annealing in neutral ambients like N_2 (Deal *et al* 1967, Revesz 1967, Montillo and Balk 1971). Again, the question regarding the model is undecided: structural disorder (Revesz *et al* 1967) and oxygen vacancies (Fowkes and Hess 1973) have been proposed, but intrinsic levels might also be considered (DiStefano and Eastman 1971a, Revesz 1973).

The creation of fast states and positive oxide charge under positive and negative bias ('slow trapping instability') has been further investigated in recent studies (Rossel *et al* 1970, Goetzberger *et al* 1973, Osburn and Chou 1973). The increase is larger for larger initial state densities, at least under negative bias stress; it is roughly proportional to the field and to the logarithm of the stressing time. Osburn and Chou (1973) found that the surface state generation under positive bias and the accompanying increase in positive oxide charge was most likely a localized phenomenon, that would lead to a strong reduction of the surface barrier, increased current injection and dielectric breakdown. Also in this case the detailed degradation mechanism is not clear at the present time. Oxide films grown in O_2-HCl mixtures show enhanced instability under high negative stress; the amount of surface charge induced during high field stressing appears to increase with increasing HCl content of the O_2 oxidation ambient at a given temperature (van der Meulen *et al* 1973). However, the O_2–HCl prepared films appear to have excellent breakdown strength (Osburn and Chou 1973).

3.3. Phosphosilicate glass

Phosphosilicate glass (PSG) films for FET gate applications are generally formed by diffusing P_2O_5 from the gas phase into thermally grown SiO_2 layers. Since this material has been used for over ten years a considerable level of sophistication has been reached in its technology. The phase diagram of the SiO_2–P_2O_5 system (Kooi 1964, Snow and Deal 1966, Eldridge and Balk 1968) shows that for deposition temperatures between 800 and 1100 °C a concentration range from 10 to 21 mole % P_2O_5 is accessible. Lower P_2O_5 concentrations may be realized by subsequent annealing treatments (Balk and Eldridge 1969). Since the latter compositions are in the two phase region between liquidus and solidus, it is possible that these films consist of two phases.

The addition of P_2O_5 to SiO_2 causes a moderate increase in the dielectric constant—from 3·82 to a value between 4·0 and 4·1 for a glass containing 16 mole % P_2O_5 (Snow and Deal 1966). In addition, the glasses can be polarized by an electric field (Snow and Deal 1966, Eldridge *et al* 1969). This effect is rather strongly dependent on the P_2O_5 content of the films, which are thought to consist of a silicate network with some of the SiO_4 tetrahedra substituted by PO_4. The motion of a non-bridging O ion, associated with every other PO_4 group, would be responsible for the polarizability of the films.

The original reason for using PSG films is their ability to retard Na^+ ion drift and the accompanying charges in FET threshold. PSG acts as a getter for this species where the maximum amount of Na being trapped is directly related to the P_2O_5 concentration. The films appear to have a distribution of trap energies. Thus the Na^+ ions will be more firmly held for larger P_2O_5 to Na^+ ratios, since they result in occupation of only the deeper traps. The trapping has been explained as being coulombic in nature, and taking place at the negatively charged non-bridging oxygen ions, that also play a role in the film polarization (Eldridge and Kerr 1971).

PSG layers exhibit a dielectric strength which is at least as good as that of SiO_2 films (Snow and Deal 1966). In addition, they show better long term resistance against breakdown (Osburn and Chou 1973). This favourable behaviour is probably related to the fact that the films are formed as liquids in an atmosphere which is strongly undersaturated in P_2O_5 compared to the pure material. There is no risk of forming particulate matter in the gas phase, as in other pyrolytic deposition processes. Thus, the liquid phase will solidify to a glassy film with very uniform properties. Lowering of the barrier for electron injection and local field enhancement due to Na^+ ion drift are also strongly retarded in PSG, which eliminates another possible cause of breakdown.

3.4. Silicon nitride

Silicon nitride layers have been prepared by a variety of methods, of which the high temperature (typically between 650 and 1000 °C) vapour phase reaction between SiH_4 (Doo *et al* 1966) or also $SiCl_4$ (Chu *et al* 1967) and NH_3 in a N_2 or H_2 carrier appears to be most widely used. The reason for the preference for the SiH_4 process lies most likely in the relative convenience of this technique. The data presented in the following are for films prepared by this process, unless stated otherwise. Excellent film properties have also been claimed for films deposited at temperatures at 500 °C or below by RF flow discharge reaction between SiH_4 and N_2 (Gereth and Scherber 1972). The electrical characteristics of the vapour deposited films have been found to be very dependent on the details of the deposition process, like gas phase reactant concentrations, carrier gas and deposition temperature. This probably implies that the composition and structure of the Si_3N_4 films may be varied more readily and that impurities may be incorporated more easily in Si_3N_4 than in SiO_2 films obtained by thermal oxidation.

Most authors appear to agree that pyrolytically prepared silicon nitride films are amorphous when deposited on properly cleaned substrates (Grieco *et al* 1968, Dalton and Drobek 1968, Chaudhari *et al* 1973). The material has most likely $Si-(Si_xN_y)$ tetrahedra ($x+y=4$) as the basic structural units, even when the composition does not correspond to the stoichiometric formula Si_3N_4 (Philipp 1973). For NH_3/SiH_4 ratios less than 10 in the deposition system the material is probably Si-rich, for higher ratios the composition approaches the stoichiometric value (Taft 1971), except for a trace amount of H (D M Brown *et al* 1968). The index of refraction for the stoichiometric composition is 2·00–2·03; the lower values observed for very high NH_3/SiH_4 ratios may be due to H (Taft 1971) or O in the films. This latter impurity, which is readily taken up, might originate from the NH_3, which is difficult to purify. The observed values for the dielectric constant range from 6·6 to 7·4 (C M Osburn 1972 unpublished, G A Brown *et al* 1968, D M Brown *et al* 1968); they will increase for very low

NH_3/SiH_4 ratios (Doo *et al* 1968) and decrease at low deposition temperatures (Duffy and Kern 1970a). Goodman (1968) has derived a band gap of 5·1 eV from photoemission measurements.

The current transport in Si_3N_4 is bulk controlled. At room temperature the field assisted thermal excitation of trapped electrons into the conduction band (Poole–Frenkel effect) dominates the conduction process (Sze 1967) (figure 3). The conductivity of the film is strongly dependent on deposition temperature and NH_3/SiH_4 ratio. Figure 4 shows an example of these dependencies, measured at room temperature for a field of $4·84 \times 10^6$ V cm^{-1} (Sewell *et al* 1970), for a SiH_4–NH_3 deposited film with Ar as a diluent (Wegener and Sewell 1969). It may be seen that higher resistivities are obtained for higher deposition temperatures and higher NH_3/SiH_4 ratios. The same trends were reported by Naber and Lockwood (1973) for a N_2 diluent. Incorporation of some oxygen in the film, using N_2O, will increase the resistivity (Wegener and Sewell 1969). Goodman *et al* (1970) observed the opposite dependence of

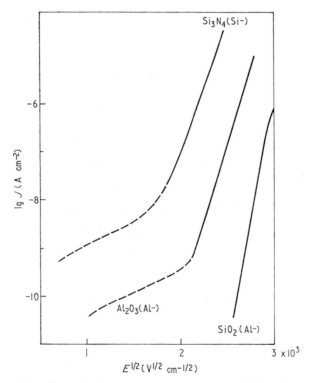

Figure 3. Field dependence of current in Si_3N_4, Al_2O_3 and SiO_2 (Sze 1967, Tsujide *et al* 1970, Osburn and Weitzman 1972) (steady state). Dotted parts of Si_3N_4 and Al_2O_3 curves were probably measured during initial transient (C M Osburn 1973 unpublished).

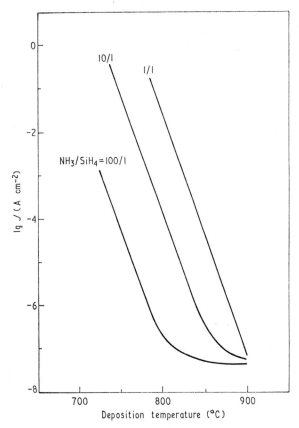

Figure 4. Current in Si_3N_4 films at $4\cdot84 \times 10^6\ V\,cm^{-1}$ for different deposition temperatures in the SiH_4–NH_3–H_2 system. Results for three NH_3/SiH_4 ratios are shown (Sewell *et al* 1970.)

resistivity on deposition temperature when using H_2 as a carrier. An increase in resistivity with NH_3/SiH_4 ratio (saturating for ratios greater than 10) was found also for this case (G A Brown *et al* 1968). For SiH_4–NH_3–H_2 deposited samples an increase in resistivity can be realized by subsequent N_2 or Ar anneal at 800–$1000\ °C$ (Lundström and Svensson 1972a).

Si_3N_4 films exhibit a very high value of dielectric strength: approximately $10^7\ V\,cm^{-1}$ (Sze 1967, G A Brown *et al* 1968). Since the material is a better conductor than SiO_2, the breakdown mechanism is thermal, that is to say breakdown takes place via thermal runaway (Sze 1967). Films deposited using a N_2 carrier show lower dielectric strength and more instances of excessive localized conduction than those deposited in H_2, whereas the films formed at the highest NH_3/SiH_4 ratios exhibited the lowest pinhole densities (Duffy and Kern 1970a). From the temperature dependence of the conductivity one may infer a trap depth of $1\cdot1$–$1\cdot2$ eV below the conduction band from the work of Sze (1967) and that of Chaudhari *et al* (1973). Kendall (1968) reports trap depths between $0\cdot50$ and $0\cdot85$ eV below the conduction band from thermally stimulated current measurements. It appears that lower resistivity material contains higher trap densities. Material with a high density of trapping centres may also have a reduced bandgap.

Si_3N_4 is an excellent barrier against Na^+ drift (Dalton and Drobek 1968). On the other hand, at temperatures comparable to its deposition temperature it will act as a getter for Na^+ ions, which will be released by annealing in moist atmospheres at 450 °C (Burgess *et al* 1969).

The drawbacks of Si_3N_4 are its relatively large conductivity (figure 3) and the fact that it can readily be etched only in strong HF solution and in hot H_3PO_4. Thus the attempt to wed the favourable properties of SiO_2 to those of Si_3N_4 has been made by preparing silicon oxynitrides. Indeed, the full range of solid solutions of SiO_2 in Si_3N_4 is experimentally accessible. They can be most conveniently prepared from SiH_4 with a large excess of NH_3 and with NO as the oxidizing agent, although other oxidants may be used as well (D M Brown *et al* 1968, Tombs *et al* 1969). Oxygen-rich compositions may also be obtained by the SiH_4–NO reaction (Rand and Roberts 1973). The amorphous films have band gaps between those of Si_3N_4 and SiO_2, increasing somewhat more slowly with increasing SiO_2 content for the nitride-rich compositions than for the oxide-rich ones (D M Brown *et al* 1968). Philipp (1973) has concluded that the basic structural units are again Si tetrahedra, in this case of the type $Si–(Si_xN_yO_z)$ with $x+y+z=4$, the atoms being statistically distributed for any given composition. Other properties such as etch rates in HF, dielectric constants and the ability to retard Na^+ ion motion are also a compromise between those of the parent compounds. The films prepared without NH_3 tend to be somewhat better Na^+ barriers than those of the same nominal composition made with NH_3, but the protection afforded against Na^+ drift is always ample for films with compositions around SiON. The effect of the method of preparation also shows up in the fact that compositions made with NH_3 are not exactly Si^{4+}, O^{2-}, N^{3-} polymers, and that they contain some H which may be bonded to N (Rand and Roberts 1973).

3.5. Aluminium oxide

Al_2O_3 films for electronic applications may be prepared by pyrolytic decomposition of Al-isopropoxide in N_2 or O_2, or by reacting trimethyl-Al with NO or O_2 atmospheres at temperatures of at least 400 °C (Aboaf 1967, Duffy and Kern 1970b, Hall and Robinette 1971). The trimethyl–NO process has been used up to temperatures of at least 800 °C. An alternate method is the reaction of $AlCl_3$ or $AlBr_3$ with a gas mixture which will produce H_2O at the reaction temperature of 400 °C or higher (CO_2+H_2, $NO+H_2$) (Tung and Caffrey 1965, Iida and Tsujide 1972, Aboaf *et al* 1973), but more typically 800 to 900 °C in this case. The reason for using the H_2O producing mixture instead of H_2O vapour directly is that this approach allows thorough mixing of all reagents without incurring the danger of forming Al_2O_3 already in the cold parts of the apparatus. Material deposited at 800 °C and above is crystalline. Deposition in the 800–900 °C range yields the kappa (Nigh 1969) or the gamma phase (Chou and Tsang 1971) or both (Iida and Tsujide 1972); at 1000 °C the theta and alpha phases are obtained (Iida and Tsujide 1972). The grain size is substrate dependent: for a 900 °C deposit it is less than 100 Å on thermally grown SiO_2, and of the order of 3 μm on Si (Nigh 1969). Below 800 °C the deposits are amorphous.

Films prepared with the $AlCl_3$–CO_2–H_2 system below 800 °C contain uniformly distributed Cl (at least 1%, more for lower temperatures, Colby 1969, Kamoshida *et al* 1971), which cannot be completely removed by subsequent annealing at temperatures as high as 900 °C (Kamoshida *et al* 1972). The lowest temperature deposits (below 550 °C) contain also OH groups (Iida and Tsujide 1972). Material from the Al-isopropoxide process is close to stoichiometry, but contains approximately 3% carbon (Colby 1969). Since all films obtained using the $AlCl_3$–CO_2–H_2 system are O deficient (particularly at the lower deposition temperatures, Colby 1969) it stands to reason that one would try to remove this deficiency by annealing in O_2. Pappis and Kingery (1961) have been able to change the conduction of single and polycrystalline Al_2O_3 from n-type (O deficient material, with O^{2-} vacancies or Al^{3+} interstitials) to p-type (metal deficient material, with Al^{3+} vacancies) by O_2 annealing above 1300 °C. However, an anneal at 1100 °C does not even seem to change the Al/O ratio to its stoichiometric value (Aboaf *et al* 1973). In addition, O_2 diffuses through the films and will oxidize a Si substrate under an amorphous Al_2O_3 film at a similar rate as under a SiO_2 layer of equal thickness, but much slower under a crystalline film (Kamoshida *et al* 1972, Iida and Tsujide 1972).

Given the problem that pyrolytic Al_2O_3 films, which have a low impurity content and approach stoichiometry, are polycrystalline, whereas the amorphous deposits are impure and often far from stoichiometry, it would be interesting to consider material obtained by other methods. Unfortunately, no analytical data are available on films prepared by sputtering or plasma oxidation.

Al_2O_3 has an optical bandgap of 8·7 eV (Laufer *et al* 1965). Trapping states have been found by several workers at the following energies below the conduction band: around 1·0 eV (thermoluminescence, Mitchell 1968), around 2·8 eV (electroluminescence, Filaretov *et al* 1968), a distribution extending throughout the upper part of the Al_2O_3 bandgap (optical depopulation, Mehta *et al* 1972, Harari and Royce 1973). Dielectric constants from 7·6 to 9·5 have been reported for polycrystalline material, those from 9·0 to 9·5 being the more likely ones (Nigh 1969, Aboaf *et al* 1973); a value of 8·5 is given for amorphous high temperature films (Hall and Robinette 1971) and for sputtered films (Pratt 1969, Salama 1970).

Conduction in polycrystalline Al_2O_3 appears to be barrier limited since it depends on the electrode material (Walden and Strain 1970, Walden 1972). Amorphous films show higher conductivity (Tsujide and Iida 1972b). There is an indication that traps participate in the current transport through the barrier. Salama (1971) claims that sputtered films exhibit Poole–Frenkel conduction for higher fields at room temperature. The material is polarizable. The effect is small and recoverable (Gnadinger 1971, Walden and Strain 1970). The breakdown strength of the gamma-Al_2O_3 layers is reported to be between 6 and 10×10^6 V cm^{-1} (Nigh 1969, Tsujide *et al* 1970). There are indications that the breakdown takes place by a thermal mechanism. Low-temperature pyrolytic films appear to break down at 7×10^6 V cm^{-1} (Carnes and Duffy 1971), but they do show considerable numbers of weak spots and areas with excessive localized conduction (Duffy *et al* 1971), which makes this material rather undesirable as a gate insulator. For sputtered Al_2O_3 a breakdown strength up to 8×10^6 V cm^{-1} has been found (Pratt 1969, Salama 1970).

Al_2O_3 retards Na^+ ion drift (Nigh *et al* 1967, Salama 1970). However, even at 300 °C small quantities of Na will penetrate into the entire film by thermal diffusion (Tung and Caffrey 1969).

4. Layered FET gate insulators

4.1. General requirements

An acceptable gate insulator has to fulfil a number of essential requirements. The system should allow elimination of the fast states at the silicon–insulator interface. It should be possible to remove the fixed charge, or set it at a desirable predetermined value. Prolonged electrical stressing should leave the silicon surface properties unaffected. The films should exhibit a low incidence of low-voltage breakdown and good wear-out characteristics, that is they should reduce their breakdown strength under electro-thermal stress only very slowly.

Because of the favourable properties of the Si–SiO_2 interface the layer in contact with the Si is generally chosen to be thermally grown SiO_2. To prevent instability due to tunnelling through the SiO_2 layer or into its conduction band, the thickness should be at least 50 Å and more if the next layer is thicker and has a high dielectric constant. In this section we will discuss the available data for the SiO_2–PSG, SiO_2–Si_3N_4 and SiO_2–Al_2O_3 layers. The focus will be on the stability of the systems.

4.2. SiO$_2$–PSG

First introduced because of its ability to block Na^+ ion drift, this insulator is particularly useful for application in n-channel FET technology. PSG compositions for this purpose contain between 1 and 7 mole % P_2O_5. At these low concentrations one would not expect any drastic effects on the band gap. Conductivity data are not available.

As prepared, the M–SiO$_2$–PSG-Si system exhibits a more negative flatband voltage than the M–SiO$_2$–Si system with the same insulator thickness. Kalter *et al* (1971) conclude from etch-down experiments that the effect may be due to a dipolar layer at the SiO$_2$–PSG interface but that the metal–insulator barrier also plays a role. They found that such interfacial dipoles are rather common in layered dielectrics. For the SiO$_2$–PSG combination investigated by Kalter the change in flatband voltage was -0.5 V. The magnitude of the shift is dependent on glass composition, but the exact relationship and the possible effect of the thickness of the layers is not known. For layers such as are used in FET gates the change is approximately -0.1 V (P Balk 1969 unpublished).

PSG layers retard Na$^+$ ion motion by a gettering mechanism; the ions will distribute themselves throughout the film. Their effect on V_{FB} will thus be proportional to the total amount of contamination and to the PSG thickness (Kaplan and Lowe 1971). Furthermore, the level of protection provided is directly related to the total amount of P$_2$O$_5$ in the film. Finally, the polarizability of PSG films is a quadratic function of the P$_2$O$_5$ concentration. These considerations allow one to determine the required thickness and P$_2$O$_5$ concentration of the PSG film. Taking both polarization and Na$^+$ charge into account, it appears that a threshold stability of 0·1 V for a 1000 Å gate, under a 2×10^6 V cm^{-1} stress up to 140 °C for 10 years is only obtainable for Na levels of the order of 10^{11} cm^{-2} (Balk and Eldridge 1969, Kaplan and Lowe 1971). Thus the protection afforded by this approach is limited.

Removal of interface states presents no problem, since PSG appears to be readily permeable to H$_2$. Oxide charge is eliminated by high temperature N$_2$ annealing, as in single SiO$_2$ films. SiO$_2$–PSG layers are very effective in reducing low-voltage breakdown (Chou and Eldridge 1970, Osburn and Ormond 1972) and in slowing down wear-out (Osburn and Chou 1973). The cause of this lies probably in the fact that upon formation the liquid PSG layer will heal defects and trap Na$^+$ ions.

4.3. *SiO$_2$–Si$_3$N$_4$*

MNOS capacitors as prepared exhibit a larger negative V_{FB} value than comparable MOS capacitors. Cricchi *et al* (1972) ascribe this to a positive charge connected to the nitride–oxide interface. They find that the magnitude of the effect depends on nitride deposition rate and temperature. For this system Kalter *et al* (1971) did not find a step in their etch-down C–V curves, but also in this case the effects of electrode barrier and insulator–insulator barrier are hard to separate.

As stated before, Si$_3$N$_4$ is an excellent Na$^+$ ion barrier. However, in order to prepare stable MNOS structures, one first has to grow stable (ie Na free) SiO$_2$ films. This may cause some difficulties in the SiH$_4$–NH$_4$–H$_2$ system (MacKenna and Kodama 1972), where the H$_2$ seems to mobilize the Na (Burgess and Donega 1969). This problem can be eliminated by adding HCl to the gas phase during or after the oxidation, or during the nitride deposition. The Na present in the SiO_2 will probably become at least partially trapped at the Si_3N_4–SiO_2 interface (Burgess *et al* 1969).

A band diagram for the Si–SiO_2–Si_3N_4–Al system is given in figure 5(*a*). It is composed of data for the bandgap of SiO_2 (DiStefano and Eastman 1971a), the electron and hole barriers for Si–Si_3N_4 (Goodman 1968), the energy difference between the Si valence band and the SiO_2 conduction band (Williams 1965), the band gap of Si (Grove 1967), the electron barrier for Al–Si_3N_4 (Goodman 1968) and the work function of Al (Riviere 1957). Outstanding features of this diagram are the relatively low electron barrier at the Al contact and the SiO_2–Si_3N_4 barrier (which has not been measured directly). This diagram along with the relatively large conductivity difference between SiO_2 and Si_3N_4

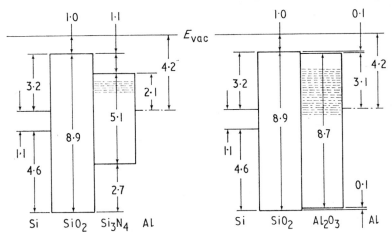

Figure 5. Tentative band diagrams for the Si–SiO_2–Si_3N_4–Al and Si–SiO_2–Al_2O_3–Al systems.

suggests that particularly under negative bias on the Al electrode a negative space charge may build up. This has indeed been observed (Curry and Nigh 1970). It is peculiar that the charge accumulation is somewhat electrode dependent, since it has been reported that the conduction in Si_3N_4 is bulk limited. Curry and Nigh's data for capacitors with 1000 Å of SiO_2 and 500 Å of Si_3N_4, deposited with the $SiCl_4$–NH_3 process at 900 °C indicate that for a nitride field of 1×10^6 V cm^{-1} (ie an oxide field of approximately 1.8×10^6 V cm^{-1}) at room temperature a total shift of 0·01 V can be expected. Data for higher temperatures are not available. Of course, one can reduce the magnitude of the V_{FB} shift by reducing the thickness of the Si_3N_4 film.

At lower temperature (below 600 °C), Si_3N_4 is impervious to H_2. It is therefore difficult to eliminate interface states completely after the nitride film has been deposited, since H_2 will be effective only below 600 °C. This is an argument in favour of the SiH_4–NH_3–H_2 process, which will probably saturate the SiO_2 layer with H_2 before sealing it off (Deal *et al* 1969, MacKenna and Blanchard 1972). The SiH_4–NH_3–N_2 process yields somewhat higher interface state densities (Duffy and Kern 1970a).

4.4. SiO_2–Al_2O_3

One of the original attractions of this system was the positive shift in V_{FB} (of the order of 1 V) for MAOS with respect to comparable MOS capacitors (Nigh *et al* 1967), which allows fabrication of p-channel FET devices with small threshold voltage or enhancement mode n-channel devices. The magnitude of this shift depends on the method by which the films were prepared; it is different for Al_2O_3 films deposited under differing conditions, it can be affected by annealing but it is independent of the thicknesses of the SiO_2 and Al_2O_3 layers (Nigh 1969, Zerbst 1969, Nishimatsu *et al* 1969, Hart and Walsh 1970). Here again, there is some difficulty in distinguishing between insulator–insulator interfacial effects and the metal–interface barrier (Nigh 1969, Kalter *et al* 1971). However, after a careful analysis of the data it appears that the observations can be explained by the presence of a negative charge at the Al_2O_3–SiO_2 interface. The charge is independent of the thickness of the Al_2O_3, and inversely proportional to the SiO_2 thickness (Labuda *et al* 1971, Aboaf *et al* 1973). It is thought that during the deposition the system is sufficiently conductive to maintain thermal equilibrium (ie equal Fermi levels) between Si and Al_2O_3. The voltage required to realize this will be determined by the bulk Fermi levels of the Si and Al_2O_3 films, giving rise to a constant voltage drop across the SiO_2, and thus to the observed dependence of the charge on SiO_2 thickness. This situation will be frozen in upon cooling.

Al_2O_3 retards mobile ion drift in layered structures with SiO_2 (Nigh *et al* 1967, Hart and Walsh 1970). When Na^+ drifts through the film, it becomes trapped at the SiO_2–Al_2O_3 interface (Abbots and Kamins 1970). The SiO_2 film under Al_2O_3 deposited from $AlCl_3$ will most likely be stable due to the action of the HCl developed in the reaction. Since this reaction is generally carried out in excess H_2, one would not expect to find substantial concentrations of interface states. However, Nakagiri and Wada (1972) observed a substantial reduction in the field effect mobility for FET devices with MAOS gates compared to MOS gates, suggesting significant concentrations of interface states. This reduction disappears only partially upon a 500 °C H_2 anneal.

The band diagram for the Si–SiO_2–Al_2O_3–Al system is also shown in figure 5. Again, data for different systems had to be used: the band gaps of Si (Grove 1967), SiO_2 (DiStefano and Eastman 1971a) and Al_2O_3 (Laufer *et al* 1965), the energy difference between the conduction bands of Si and Al_2O_3 (Szydlo and Poirier 1971), the barrier energy between Al and Al_2O_3 (Szydlo and Poirier 1971), the energy difference between the Si valence band and the Al_2O_3 conduction band (Williams 1965) and the work function of Al (Riviere 1957). The different trapping centres, which have been reported for Al_2O_3 (see section on Al_2O_3) are also indicated.

The barrier between Al electrode and Al_2O_3 is quite respectable. However, as mentioned before, charge injection from the metal takes place most likely with the aid of the traps. Thus, here again V_{FB} will shift particularly under negative bias on the metal. At temperatures up to 200 °C one observes a V_{FB} shift which is symmetric with the applied field, and which behaves like a polarization affect (ΔV_{FB} positive for negative bias, negative for positive bias). Iida *et al* (1972) assume that the polarization effect occurs in an aluminosilicate layer at the SiO_2–Al_2O_3 interface. Under positive bias an additional mechanism plays a role, which again decreases V_{FB}. It is thought to be caused by mobile charges drifting from the metal–Al_2O_3 interface into the insulator (Gnadinger 1971, Gnadinger and Rosenzweig 1973). At higher temperatures and negative bias a slow trapping effect also becomes visible. Stressing data for 1000 Å of SiO_2/500 Å of Al_2O_3 capacitors under 10 V negative bias (ie an SiO_2 field of 0.8×10^6 V cm^{-1}) indicate threshold shifts of below 0·1 V in 40 years at 85 °C (Lampi and Labuda 1972). For positive bias a shift of 0·35 V would be expected under the same conditions. Thus the SiO_2–Al_2O_3 double insulator seems to be more favourable for p-channel than for n-channel devices. Nishimatsu *et al* (1973) have recently found that the polarizability of the Al_2O_3 films may be reduced by a factor of 4 by purification of the $AlCl_3$ source material.

The breakdown strength of the layered dielectric is dependent on the quality of the SiO_2 film underneath (Hart and Walsh 1970). Because of the polycrystalline nature of Al_2O_3 one would expect that there is a minimum practical thickness of the Al_2O_3, comparable with the grain size of the material, below which the film would not be continuous any more.

5. Non-volatile memory devices

5.1. Charge injection and storage in layered structures

The potential instability of the metal–insulator–silicon structure with layered dielectric gate can be turned to advantage and utilized to obtain storage devices with bistable threshold. The most attractive way to realize this concept is by employing an MIOS configuration where the insulator I is virtually non-conducting compared to the SiO_2 under application of a bias voltage. This behaviour may be promoted by choosing for I a material which, in addition to having a low conductivity, also has a relatively high dielectric constant. According to equation 2 the field in the SiO_2 will be higher than that in the other insulator. Therefore, given the experimental dependence of the conductance on the field, the current in insulator I will be small compared to that in SiO_2 and charge will accumulate in traps near the I–O interface. The trapped charge will gradually reduce the field across the SiO_2 layer and increase that over I; consequently the current in the SiO_2 will decrease and that in I will increase until a steady state

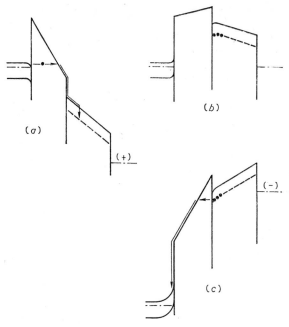

Figure 6. Charging and discharging in MIOS memory devices with thick SiO_2 layer for (*a*) positive bias, Fowler–Nordheim tunnelling, (*b*) positive bias removed and (*c*) negative bias.

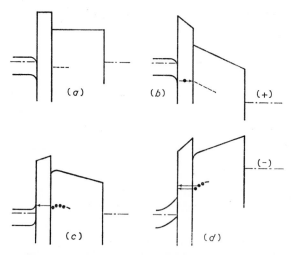

Figure 7. Charging and discharging in MIOS memory devices by direct tunnelling: (*a*) before charging, (*b*) positive bias, (*c*) positive bias removed, and (*d*) negative bias.

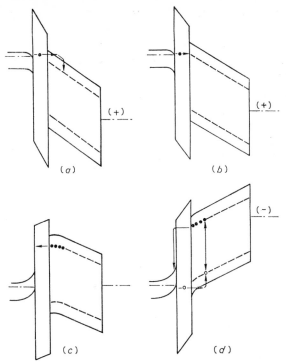

Figure 8. Charging and discharging in MIOS devices with very thin SiO$_2$ film: (*a*) positive bias, modified Fowler–Nordheim tunnelling, (*b*) small positive bias: direct tunnelling, (*c*) positive bias removed and (*d*) negative bias.

is reached. At this point the total amount of charge collected, and thus the V_{FB} of the system, will reach a saturation value. This situation has been depicted in figure 6, where it has been assumed that the SiO$_2$ current is determined by Fowler–Nordheim tunnelling. Upon removal of the gate voltage the field due to the trapped charge is generally too small to cause any reverse current, and the charge will be maintained indefinitely. It will only tunnel back when applying a voltage of opposite sign (Frohman-Bentchkowsky and Lenzlinger 1969).

Devices with very thin SiO$_2$ films (below 40 Å in thickness) require smaller voltages for switching. Here two other mechanisms have been proposed. In the first (figure 7), electrons tunnel through the SiO$_2$ layer directly into the traps in I that were originally aligned with the forbidden band of silicon (Wallmark and Scott 1969, Ross and Wallmark 1969, Dorda and Pulver 1970). The second model (figure 8) assumes tunnelling from the Si conduction band into the conduction band of I (modified Fowler–Nordheim process) where the electrons subsequently move towards the metal electrode and become trapped (Svensson and Lundström 1970). Only for smaller fields would direct tunnelling from the

Si conduction band into the traps be possible. This process would terminate as soon as the field increases and the trap drops below the conduction band of Si. With these thin SiO_2 films it is possible that in the absence of an external bias a large charge will leak back to the silicon by its own field. Under negative bias holes may be injected from the silicon valence band into the insulator valence band. These holes will become trapped and may eventually recombine with electrons that are already present (Lundström and Svensson 1972a). In addition, the trapped electrons may tunnel back into the Si.

Thus both in MIOS devices with a thick SiO_2 layer and in those with a very thin one, two states of charge are obtained. The difference between the V_{FB} (or V_T) values in these situations (the so-called 'window') is a function of the oxide thickness and of the amplitudes and the lengths of the applied bias pulses for a given effective total insulator thickness (capacitance). All models indicate that a minimum field in the oxide (and thus a minimum applied voltage) is required to start the charging process. Voltages below this minimum value can be used to interrogate the state of the device. Above that value, at a given window size and insulator thickness shorter charging times are required for larger fields. Thus a trade-off between length and amplitude of the charging (writing) pulse appears possible. A fundamental limitation on pulse amplitude is given by the breakdown voltage of the gate structure. There is also the practical consideration that high voltages are generally incompatible with the operation of the supporting MOSFET circuitry on the memory chip.

If for insulator I a material is chosen with a smaller electron affinity (ie a smaller $I–SiO_2$ electron barrier) a larger oxide field will be required to obtain a given electron current by modified Fowler–Nordheim tunnelling. At a sufficient density of trapping states in I direct tunnelling from the silicon into the traps might make a more important contribution to the charging in this case.

One can only distinguish between the two aforementioned models for very thin SiO_2 devices on the basis of a detailed analysis of their charging and discharging behaviour in dependence on time, field and oxide thickness. Such an analysis has been given for the second model in a series of papers by Svensson (1971), Lundström and Svensson (1972a, b), Carlstedt and Svensson (1972) and Lundkvist *et al* (1973). A straightforward condition for the correctness of the first model is the following: a minimum field across the SiO_2 will be required to line up those states, which lie in the energy range of the silicon band gap in the field-free condition, with the valence and conduction band edges of the silicon in order to charge or discharge them. This implies that the difference between the voltage drops across the oxide for these conditions should be 1.1 V for states located at the $SiO_2–I$ interface, and even less for those deeper into insulator I (Gordon and Johnson 1973).

It should be pointed out here that the modified Fowler–Nordheim tunnelling model implies that very substantial fields in the oxide (over 10^7 V cm^{-1}) would be required to switch the state of such memory devices at a high speed. As

discussed earlier, one may expect to observe slow trapping effects and con-
comitant changes in the $Si–SiO_2$ interfacial region for such fields. These
effects (increases in the densities of surface charge centres and of fast interface
states) would lead to a shift in the 'window' position and a lowering of the transconductance of the memory device, and thus present serious reliability problems.

In the next sections we will review some experimental data for MNOS and
MAOS non-volatile memory devices regarding the points discussed above.

5.2. MNOS devices

The desirability of good charge retention and fast charging through a thin SiO_2
film requires a high resistivity nitride film. Thus, when depositing in a H_2
atmosphere, a low deposition temperature is chosen (700 °C, Goodman *et al*
1970) eventually followed by an Ar or N_2 anneal at 800–1000 °C to reduce the
oxide charge (Lundström and Svensson 1972a, b). The deposition in N_2 is
probably best carried out at a higher temperature (850 °C) to obtain this effect
(Naber and Lockwood 1973). It is not known what interface state densities and
transconductances are obtained on devices with nitride layers grown in H_2-poor
ambients. There appears to be agreement now that for devices with very thin
SiO_2 films charge leakage takes place by back tunnelling through the oxide layer
(Lundström and Svensson 1972a, Naber and Lockwood 1973, Yun 1973).
Given the interest in obtaining low switching voltages, it is this type of MNOS
device which has been studied most.

Because of the dominant role that the SiO_2 layer plays in the current transport to and from the traps and the strong dependence of the tunnel currents
through the oxide on the thickness of the layer good control of this parameter
is imperative. The position and width of the window for a given bias pulse
amplitude depend strongly on SiO_2 thickness, and so does the rate of loss of
stored charge (Ross *et al* 1970, Chou *et al* 1972, Kobayashi 1972). Thicknesses
of about 20 Å appear to be favourable, because the device properties are not
extremely sensitive to oxide thickness in this range, the rate of charging relatively fast and the required voltages relatively moderate.

It appears that the charging and discharging characteristics of MNOS devices
do not only qualitatively follow the modified Fowler–Nordheim model for
electron and hole tunnelling into the Si_3N_4 conduction and valence bands
respectively (Kobayashi and Ohta 1972, Gordon and Johnson 1973) but exhibit
all predicted dependencies for this case (Lundström and Svensson 1972a,
Lundkvist *et al* 1973). In addition, direct tunnelling from the silicon conduction band into nitride traps had to be assumed for small fields during charging
(Lundström and Svensson 1972a, Maes and van Overstraeten 1973). The reverse
process plays a role as part of the discharging process. Further support to
the model is given by Gordon and Johnson's (1973) finding that devices in Ross
et al (1970) and also their own thin SiO_2 devices show a larger difference between
the minimum charging and discharging voltages than would be compatible with

F

a mechanism in which tunnelling into and from states at energies of the silicon band gap would play a role. Conti and Vanin's (1972) single example of a capacitor that does seem to fit this model may be fortuitous because of the large uncertainty in their SiO_2 thickness measurements. White and Cricchi (1972) believe the direct tunnelling mechanism to apply, but their data do not permit using the above mentioned criterion. They claim that the charge retention times are being determined by direct tunnelling from the traps into $Si-SiO_2$ interface states. The modified Fowler–Nordheim tunnelling model implies that flow of electrons in the nitride conduction band is a prerequisite for charging the traps. Gordon and Johnson (1973) observed that appreciable charging only occurred for voltages at which a significant DC flows through the device. Also, Yun and Arnett (1972) established that the centroid of the trapped charge for devices with a 25 Å SiO_2 film may be as much as 100 Å into the nitride.

Using devices with 25 Å of SiO_2 and 335 Å of Si_3N_4 White and Cricchi (1972) obtain a write/erase pulse width of 0·1 μs and a 3 V threshold window for oxide fields of $1·2 \times 10^7\,V\,cm^{-1}$, which amounts to bias voltages of $+17·5$ and 25 V. If longer writing times are permitted, these voltages may be considerably reduced.

Under such enormous electrical stress the SiO_2 and $Si-SiO_2$ interface properties will gradually degrade, as discussed in the section on the $Si-SiO_2$ system. This degradation manifests itself for MNOS devices in an increase in the fast state density, a flatband shift for both charge states in negative direction, a decrease in

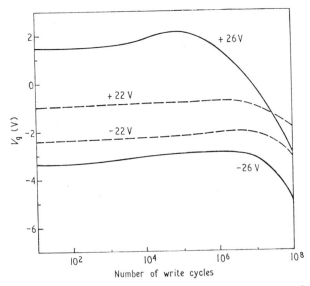

Figure 9. Change of window in MNOS memory transistor under continued cycling. Pulse width: 50 μs; Si: p-type; SiO_2: 20 Å; Si_3N_4: 406 Å; metal: Al. (Woods and Tuska 1972.)

window size and a reduced charge storage time (Ross *et al* 1971, Cricchi and Reed 1971, Woods and Tuska 1972). Thus, when using an interrogating voltage which was originally in the middle of the window, after degradation it would have moved outside. Both logical states would be indistinguishable, as neither would conduct (figure 9). Smaller amplitude pulses will reduce the window size, but move its collapse to a higher number of writing cycles. It seems possible that the degradation is related to a charging of the oxide and lowering of the barrier height, resulting in increased conductivity. These reliability problems might make the adoption of the MNOS device for RAM use difficult, and only that for electrically alterable read-only memories would remain.

Since the presence of charged centres near the conducting channel parallel to the Si–SiO$_2$ interface of an FET device appears to reduce carrier mobility (Fang and Fowler 1968) it seems logical to expect such effects in MNOS devices. Oakley and Pepper (1972) indeed found a lowering of the field effect mobility of holes in a p-channel device both for positive and negative trapped charge.

Recent publications have shown some new approaches towards writing these devices, namely by using hot electrons (Uchida *et al* 1973) or by optical stimulation (Vanstone 1972, Sewell 1973). In the latter case the MNOS device can also be made to operate in an analogue mode.

5.3. MAOS devices

Based on the hypothesis that the operation of MIOS memory devices depends on tunnelling into and from states across from the forbidden gap of the silicon one would expect to be able to obtain switching at lower voltages if a material with higher dielectric constant than Si$_3$N$_4$ ($\epsilon \simeq 7$) would be used as the outer insulator. This caused interest in using Al$_2$O$_3$ ($\epsilon \simeq 9$). The MAOS system, generally used with an Al electrode, does indeed exhibit charge storage behaviour similar to that in the MNOS device. However, when applying Gordon and Johnson's (1973) criterion that the difference between the voltage drops across the SiO$_2$ for charging and discharging could be at most 1·1 V for this model, Chou and Tsang's (1971) data show that at least 1·5 V is required for each process (ie 3 V total) for their nominally 26 Å thick SiO$_2$ films. The value of 1·5 V suggests that the charging takes place by direct tunnelling from the silicon conduction band into trapping states halfway between the conduction bands of silicon and Al$_2$O$_3$ (see figure 6). This view is also held by Ohta and Hamano (1972). Thus the question of the advantage of the MAOS above the MNOS device is not a straight-forward one any more and one can determine only experimentally which system will exhibit the lower switching voltage. Unfortunately, the data available in the literature are not sufficient to reach a conclusion on this point.

One instance where the MAOS device clearly distinguishes itself from the MNOS structure is in the pronounced tendency towards electron injection from the metal electrode into the Al$_2$O$_3$. This causes V_{FB} to shift in a positive direction not only for positive but also for negative write (or erase) voltages (Nakamura *et al* 1970, Balk and Stephany 1970, 1971, Chou and Tsang 1971), leading to the suggestion that the system could only be used for read-only memory applications. It is worth remarking that the Si_3N_4 films, which exhibit a larger steady state conductivity do not show this effect. Since it appears likely that this injection from the metal takes place into traps related to the oxygen deficiency of the material, reduction of the trap density should reduce the injection probability. It was indeed observed that annealing in O_2 suppresses the positive V_{FB} shift for negative bias voltages (Balk and Stephany 1971). The O_2 annealing also had a small effect on V_{FB} before charging (Iida 1972). Choosing a metal for which one expects a higher barrier than Al, namely Au, the injection could again be suppressed (Balk and Stephany 1971). With a liquid electrode, Tsujide (1972) did not observe injection either. It was subsequently shown by Tsujide and Iida (1972a) that one cannot explain these observations as a simple work function effect since the suppression is obtained with evaporated Pt, but not with sputtered Pt films, suggesting that the latter process creates extra trapping centres. The results of Lampi and Labuda (1972) on instabilities in MAOS structures with sputtered Pt seem to support this contention. Finally, the injection with Al electrodes was apparently not observed for sputtered Al_2O_3 films (Salama 1971) and for plasma anodized MAOS capacitors (Pappu and Boothroyd 1973).

A basic difficulty in discussing the MAOS observations is the fact that the SiO_2 and Al_2O_3 layers apparently tend to interdiffuse (Balk and Stephany 1971, Kamoshida *et al* 1972). For this reason all quantitative considerations based on the thickness of the SiO_2 layer become somewhat uncertain. Naturally, for devices with thick SiO_2 layers, exhibiting Fowler–Nordheim tunnelling, this problem is less critical. Formation of such aluminosilicate layers may also be the cause of the failure to obtain the expected value of field effect mobility upon H_2 annealing, particularly in devices with very thin SiO_2 films (Nakagiri and Wada 1972).

There is no data available on the effect of prolonged write/erase cycling on MAOS transistors. Basically, one expects the same degradation phenomena here as in MNOS devices. It has indeed been found that interface states are formed under application of negative bias voltages (Hamano and Ohta 1972).

The use in a non-volatile optical memory using photo-bias has also been proposed for the MAOS structure (Tarui *et al* 1973).

6. Summary and conclusions

The attempt to improve the stability of MOS gates against mobile ion drift by using layered dielectrics entails the basic problem that such structures are fundamentally unstable with respect to current flow. In addition, two of the three additional insulators that have been actively studied (PSG and Al_2O_3) are polarizable. Thus, all one can attempt to do is reduce the magnitude of these instabilities to an acceptable level, for example by preparing films that have the highest possible resistivity. It is clear that one may have to compromise here.

In the case of PSG films, this compromise involves the use of rather low P_2O_5 concentrations, which limits the polarization effect, but also the Na blocking capacity. It is conceivable that the compromise for Si_3N_4 may involve incorporating O into the film, thus decreasing its conductivity. Unfortunately, there are very few quantitative data in the literature on the stability of SiO_2–Si_3N_4 films. The SiO_2–Al_2O_3 combination appears to be less stable than SiO_2–PSG, but it has the advantage of having a negative charge built into the system. In principle one can improve the stability of the layered structure by making the PSG, Si_3N_4 or Al_2O_3 layer thinner. The crystallinity of Al_2O_3 films will be a handicap in this respect. For the latter case one would not expect improved breakdown behaviour due to the presence of that layer either.

The only practical layered dielectrics for obtaining very thin gate insulators, for example the equivalent of 250 Å of SiO_2, thus appear to be the SiO_2–PSG and SiO_2–Si_3N_4 combinations (eventually SiO_2–SiON). Particularly in the latter case the thickness of the outer layer (Si_3N_4) should be small compared to that of the SiO_2 film, but probably still beyond the thickness range where tunnelling could take place. Such a structure might therefore consist of 200 Å of SiO_2 and 80 Å of Si_3N_4. It appears unlikely that a further reduction in equivalent SiO_2 thickness, for example by 30%, is feasible. SiO_2–PSG leaves some room for further manoeuvring, but only at the cost of approaching single SiO_2 films!

The unstable MIOS structure offers on the other hand an attractive way to obtain non-volatile storage within the framework of the MOS technology. Even though the technology of reproducibly growing very thin SiO_2 layers is rather demanding, structures with very thin SiO_2 films appear for many applications preferable to Fowler–Nordheim tunnelling devices because they are easier to write and erase.

It has become increasingly more unlikely that the direct tunnelling mechanism (electrons tunnelling from the Si valence band through the SiO_2 into deep traps in I) is correct. This makes finding a criterion for selecting for I a material with optimal properties more complex. Thus it is not clear any more if one should expect switching at lower voltages for Al_2O_3 than for Si_3N_4.

In addition to the question of obtaining low writing voltages, which is not only important for circuit considerations but also for the possible trade-off with writing speed, there is the problem of the field in the oxide. Reasonable write/erase times for random access memories (approximately 1 μs) are apparently only attainable with SiO_2 fields over 1×10^7 V cm^{-1} for both MNOS and MAOS devices. These high fields present a fundamental problem with regard to the reliability of the device. At the present time it is not clear whether the technology can be improved to substantially retard the appearance of the collapse of the window between both logical states of the transistor.   This would only leave open its use in electrically alterable read-only memory applications. The available information does not permit making any judgments about the possible advantages or disadvantages of MAOS over MNOS structures, except that the polycrystalline nature of the Al_2O_3 film appears to be a drawback.   Considerably more materials studies are imperative in this area.

One possible way of lowering the writing field in the SiO_2 is by introducing a sufficient number of deep traps near the SiO_2–I interface.   It seems that this can be done by introducing a deep potential well in the form of a thin metal film (Kahng and Sze 1967), small metal particles (Laibowitz and Stiles 1971) or Si particles (Horiuchi 1972).   These particular schemes may not be practical, but the data suggest that the principle is correct.

Recent publications on memory structures incorporating AlN (Mirsch and Reimer 1972) and HfO_2 (Shuskus *et al* 1973) show that the possibilities are not exhausted with Si_3N_4 and Al_2O_3.   It is not unlikely that such research on high resistivity, wide bandgap materials will indicate new approaches and will increase our fundamental understanding of MIOS systems.

Acknowledgment

This paper was completed during the author's stay as a summer visitor at the IBM Research Center, Yorktown Heights, New York in July and August 1973. He is indebted to several staff members, in particular A B Fowler and C M Osburn, for discussions on the topic of this review.

References

Abbots R A and Kamins T I 1970 *Solid St. Electron.* **13** 565
Aboaf J A 1967 *J. Electrochem. Soc.* **114** 948
—— 1971 *J. Electrochem. Soc.* **118** 1370
Aboaf J A, Kerr D R and Bassous E 1973 *J. Electrochem. Soc.* **120** 1103
Balk P and Eldridge J M 1969 *Proc. IEEE* **57** 1558
Balk P and Stephany F 1970 *NTZ-Nachrtech. Z.* **23** 526
—— 1971 *J. Electrochem. Soc.* **118** 1634
Barruel F and Pfister J C 1972 *Phys. Stat. Solidi* A **10** 555
Brews J R 1973 *J. Appl. Phys.* **44** 379
Broers A N and Dennard R H 1973 *Semiconductor Silicon 1973* ed H R Huff and R R
    Burgess (Princeton: Electrochemical Society) p 830
Brown D M, Gray P V, Heumann F K, Phillipp H R and Taft E A 1968 *J. Electrochem.
    Soc.* **115** 311
Brown G A, Robinette W C Jr and Carlson H G 1968 *J. Electrochem. Soc.* **115** 948
Burgess T E, Baum J C, Fowkes F M, Holmstrom R and Shirn G H 1969 *J. Electro-
    chem. Soc.* **116** 1005
Burgess T E and Donega H M 1969 *J. Electrochem. Soc.* **116** 1313
Carlstedt L G and Svensson C M 1972 *IEEE J. Solid St. Circuits* **SC-7** 382
Carnes J E and Duffy M T 1971 *J. Appl. Phys.* **42** 4350
Chaudhari P K, Franz J M and Acker C P 1973 *J. Electrochem. Soc.* **120** 991

Chou N J, Aboaf J A, Hammer R and Crowder H P 1972 *IEEE Trans. Electron Dev.* **ED-19** 198

Chou N J and Eldridge J M 1970 *J. Electrochem. Soc.* **117** 1287

Chou N J and Tsang P J 1971 *Met. Trans.* **2** 659

Chu T L, Lee C H and Gruber G A 1967 *J. Electrochem. Soc.* **114** 717

Colby J W 1969 *Thin Film Dielectrics* ed F Vratny (New York: Electrochemical Society) p 491

Conti M and Vanin A 1972 *Thin Solid Films* **14** 211

Cricchi J R and Reed W D Jr 1971 *9th Ann. Proc. Reliability Physics, Las Vegas, 1971* (New York: IEEE) p 1

Cricchi J R, Reid P R and McLouski R M 1972 *Electrochem. Soc. Meeting, Miami Beach 1972* (Princeton: Electrochemical Society) Abstr. 167

Curry J J and Nigh H E 1970 *8th Ann. Proc. Reliability Physics, Las Vegas 1970* ed J M Morris (New York: IEEE) p 29.

Dalton J V and Drobek J 1968 *J. Electrochem. Soc.* **115** 865

Deal B E and Grove A S 1965 *J. Appl. Phys.* **36** 1370

Deal B E, MacKenna E L and Castro P L 1969 *J. Electrochem. Soc.* **116** 997

Deal B E, Sklar M, Grove A S and Snow E H 1967 *J. Electrochem. Soc.* **114** 266

Deal B E, Snow E H and Mead C A 1966 *J. Phys. Chem. Solids* **27** 1873

DiStefano T H 1971 *Appl. Phys. Lett.* **19** 280

—— 1973 *J. Appl. Phys.* **44** 527

DiStefano T H and Eastman D E 1971a *Solid State Commun.* **9** 2259

—— 1971b *Phys. Rev. Lett.* **27** 1560

Doo V Y, Kerr D R and Nichols D R 1968 *J. Electrochem. Soc.* **115** 61

Doo V Y, Nichols D R and Silvey G A 1966 *J. Electrochem. Soc.* **113** 1279

Dorda G and Pulver M 1970 *Phys. Stat. Solidi* A **1** 71

Duffy M T, Carnes J E and Richman D 1971 *Met. Trans.* **2** 667

Duffy M T and Kern W 1970a *RCA Rev.* **31** 742

—— 1970b *RCA Rev.* **31** 754

Eldridge J M and Balk P 1968 *Trans. Met. Soc. AIME* **242** 539

Eldridge J M and Kerr D R 1971 *J. Electrochem. Soc.* **118** 986

Eldridge J M, Laibowitz R M and Balk P 1969 *J. Appl. Phys.* **40** 1922

Fang F and Fowler A B 1968 *Phys. Rev.* **169** 619

Filaretov G A, Stafeev V I, Bubnov Yu Z and Chizhik T N 1968 *Sov. Phys.—Semicond.* **1** 1242

Fowkes F M and Hess D W 1973 *Appl. Phys. Lett.* **22** 377

Fritzsche C 1967 *Z. Angew. Phys.* **24** 48

Frohman-Bentchkowsky D and Lenzlinger M 1969 *J. Appl. Phys.* **40** 3307

Gereth R and Scherber W 1972 *J. Electrochem. Soc.* **119** 1248

Ghez R and van der Meulen Y J 1972 *J. Electrochem. Soc.* **119** 1100

Gnadinger A P 1971 *Electrochem. Soc. Meeting, Cleveland, 1971* (Princeton: Electrochemical Society) Abstr. 164

Gnadinger A P and Rosenzweig W 1973 *Electrochem. Soc. Meeting, Chicago 1973* (Princeton: Electrochemical Society) Abstr. 89

Goetzberger A, Heine V and Nicollian E H 1966 *Appl. Phys. Lett.* **9** 444

Goetzberger A, Lopez A D and Strain R J 1973 *J. Electrochem. Soc.* **120** 90

Goetzberger A and Nigh H E 1966 *Proc. IEEE* **54** 1454

Goodman A M 1968 *Appl. Phys. Lett.* **13** 275

Goodman A M and Breece J M 1970 *J. Electrochem. Soc.* **117** 982

Goodman A M, Ross E C and Duffy M T 1970 *RCA Rev.* **31** 342

Gordon N and Johnson W C 1973 *IEEE Trans. Electron Dev.* **ED-20** 253

Grieco M J, Worthing F L and Schwartz B 1968 *J. Electrochem. Soc.* **115** 525
Grove A 1967 *Physics and Technology of Semiconductor Devices* (New York: Wiley) p 102
Hall L H and Robinette W C 1971 *J. Electrochem. Soc.* **118** 1624
Hamano K and Ohta K 1972 *Japan J. Appl. Phys.* **11** 1053
Harari E and Royce B S H 1973 *Appl. Phys. Lett.* **22** 1973
Hart P B and Walsh P S 1970 *Electrochem. Soc. Meeting, Los Angeles 1970* (Princeton: Electrochemical Society) Abstr. 113
Hickmott T W 1972 *J. Vac. Sci. Technol.* **9** 311
—— 1973 *Appl. Phys. Lett.* **22** 267
Hofstein S R 1967 *Solid St. Electron.* **10** 657
Horiuchi M 1972 *Paper presented at Int. Electron Devices Meeting, Washington 1972*
Iida K 1972 *Japan. J. Appl. Phys.* **11** 288
Iida K and Tsujide T 1972 *Japan. J. Appl. Phys.* **11** 840
Iida K, Tsujide T and Nakagira M 1972 *Japan. J. Appl. Phys.* **11** 1153
Kahng D and Nicollian E H 1972 *Appl. Solid State Science* vol 3 ed R Wolfe (New York: Academic Press) p 1
Kahng D and Sze S M 1967 *Bell Syst. Tech. J.* **46** 1288
Kalter H, Schatorjé J J H and Kooi E 1971 *Philips. Res. Rept.* **26** 181
Kamoshida M, Mitchell I V and Mayer J W 1971 *Appl. Phys. Lett.* **18** 292
—— 1972 *J. Appl. Phys.* **43** 1717
Kaplan L H and Lowe M E 1971 *J. Electrochem. Soc.* **118** 1649
Kendall E M J 1968 *Can. J. Phys.* **46** 2509
Kerr D R, Logan J S, Burkhardt P J and Pliskin W A 1964 *IBM J. Res. Develop.* **8** 376
Kobayashi K 1972 *Japan. J. Appl. Phys.* **11** 555
Kobayashi K and Ohta K 1972 *Japan. J. Appl. Phys.* **11** 538
Kooi E 1964 *J. Electrochem. Soc.* **111** 1383
—— 1965 *Philips Res. Rept.* **20** 578
—— 1966 *Philips Res. Rept.* **21** 477
Kriegler R J 1973 *Semiconductor Silicon* 1973 ed H R Huff and R R Burgess (Princeton: Electrochemical Society) p 363
Kuhn M and Silversmith D J 1971 *J. Electrochem. Soc.* **118** 966
Kuper A B 1969 *Surface Science* **13** 172
Labuda E F, Clemens J T and Berglund C N 1971 *Paper presented at IEEE Device Res. Conf. Ann Arbor 1971*
Laibowitz R B and Stiles P J 1971 *Appl. Phys. Lett.* **18** 267
Lamb D R 1967 *Electrical Conduction Mechanisms in Thin Insulating Films* (London: Methuen) p 8
Lampi E E and Labuda E F 1972 *10th Ann. Proc. Reliability Physics, Las Vegas 1972* (New York: IEEE) p 112
Laufer A H, Prigog J A and McNesby J R 1965 *J. Opt. Soc. Am.* **55** 64
Lenzlinger M and Snow E H 1969 *J. Appl. Phys.* **40** 278
Lundkvist L, Lundström I and Svensson C 1973 *Solid St. Electron.* **16** 811
Lundström K I and Svensson C M 1972a *IEEE Trans. Electron. Devices* **ED-19** 826
—— 1972b *J. Appl. Phys.* **43** 5045
MacKenna E L and Blanchard R 1972 *Electrochem. Soc. Meeting, Miami Beach, 1972* (Princeton: Electrochemical Society) Abstr. 249
MacKenna E and Kodama P 1972 *J. Electrochem. Soc.* **119** 1094
Maes H and van Overstraeten R 1973 *Electron. Lett.* **9** 19
Mehta D A, Butler S R and Feigl F J 1972 *J. Appl. Phys.* **43** 4631
van der Meulen Y J 1972 *J. Electrochem. Soc.* **119** 530

van der Meulen Y J, Osburn C M and Ziegler J F 1973 *Electrochem. Soc. Meeting, Chicago 1973* (Princeton: Electrochemical Society) Abstr. 54

Mirsch S and Reimer H 1972 *Phys. Stat. Solidi* A **11** 631

Mitchell J P 1968 *IEEE Trans. Nucl. Sci.* **NS-15** (6) 154

Mitchell J P and Denure D G 1973 *Solid St. Electron.* **16** 825

Miura Y and Matukura Y 1966 *Japan. J. Appl. Phys.* **5** 180

Montillo F and Balk P 1971 *J. Electrochem. Soc.* **118** 1463

Naber C T and Lockwood S C 1973 *Semiconductor Silicon 1973* ed H R Huff and R R Burgess (Princeton: Electrochem. Soc.) p 401

Nakagiri M and Wada T 1972 *Japan. J. Appl. Phys.* **11** 1484

Nakamura S, Tsujide T, Igarashi R, Onada K, Wada T and Nakagiri M 1970 *IEEE J. Solid State Circuits* **SC-5** 203

Nicollian E H, Berglund C N, Schmidt P F and Andrews J M 1971 *J. Appl. Phys.* **42** 5654

Nigh H E 1969 *Int. Conf. Properties and Use of MIS Structures, Grenoble 1969* ed J Borel

Nigh H E, Stach J and Jacobs R M 1967 *IEEE Trans. Electron Devices* **ED-14** 631

Nishimatsu S, Hashimoto N, Masuhara T and Nagata M 1973 *Paper presented at 5th Int. Conf. Solid State Devices, Tokyo 1973*

Nishimatsu S, Tokuyama T and Matsushita M 1969 *Thin Film Dielectrics* ed F Vratny (New York: Electrochem. Soc.) p 338

Oakley R E and Godber G A 1972 *Thin Solid Films* **9** 287

Oakley R E and Pepper M 1972 *Phys. Lett.* **41A** 87

Ohta K and Hamano K 1972 *Japan. J. Appl. Phys.* **11** 546

Osburn C M and Chou N J 1973 *J. Electrochem. Soc.* **120** 1377

Osburn C M and Ormond D W 1972 *J. Electrochem. Soc.* **119** 597

Osburn C M and Raider S I 1973 *J. Electrochem Soc.* **120** 1369

Osburn C M and Weitzman E J 1972 *J. Electrochem. Soc.* **119** 603

Pao H C and O'Connel M 1968 *Appl. Phys. Lett.* **12** 260

Pappis J and Kingery W D 1961 *J. Am. Ceram. Soc.* **44** 459

Pappu R V and Boothroyd A R 1973 *Appl. Phys. Lett.* **22** 72

Philipp H R 1973 *J. Electrochem. Soc.* **120** 295

Powell R J and Berglund C N 1971 *J. Appl. Phys.* **42** 4390

Pratt I H 1969 *Solid St. Technol.* **12** (12) 49

Raider S I and Flitsch R 1971 *J. Electrochem. Soc.* **118** 1011

Rand M J and Roberts J F 1973 *J. Electrochem. Soc.* **120** 446

Revesz A G 1967 *Phys. Stat. Solidi* **19** 193

—— 1973 *J. Non-Crystalline Solids* **11** 309

Revesz A G and Evans R J 1969 *J. Phys. Chem. Solids* **30** 551

Revesz A G and Zaininger K H 1968 *RCA Rev.* **29** 22

Revesz A G, Zaininger K H and Evans R J 1967 *J. Phys. Chem. Solids* **28** 197

Riviere J C 1957 *Proc. Phys. Soc.* **B70** 676

Ross E C, Goodman A M and Duffy M T 1970 *RCA Rev.* **31** 467

Ross E C, Tuska J W and Duffy M T 1971 *Tech. Rep.* AFAL-TR-71-82 (Wright–Patterson Air Force Base: Air Force Avionics Lab.)

Ross E C and Wallmark J T 1969 *RCA Rev.* **30** 366

Rossel P, Martinot H and Esteve D 1970 *Solid St. Electron* **13** 425

Salama C A T 1970 *J. Electrochem. Soc.* **117** 913

—— 1971 *J. Electrochem. Soc.* **118** 1993

Sewell F A Jr 1973 *Trans. Electron Dev.* **ED-20** 563

Sewell F A Jr, Lewis E T and Wegener H A R 1970 *Tech. Rep.* AFAL-TR-70-148 (Wright–Patterson Air Force Base: Air Force Avionics Lab.)

Sewell F A Jr, Wegener H A R and Lewis E T 1969 *Appl. Phys. Lett.* **14** 45

Shuskus A J, Quinn D J and Cullen D E 1973 *Appl. Phys. Lett.* **23** 184

Simmons J G 1970 *Handbook of Thin Film Technology* ed L I Maissel and R Glang (New York: McGraw-Hill) p 14–1

Snow, E H and Deal B E 1966 *J. Electrochem. Soc.* **113** 263

Snow E H, Grove A S, Deal B E and Sah C T 1965 *J. Appl. Phys.* **36** 1664

Sprague J L, Minahan J A and Wied O J 1962 *J. Electrochem. Soc.* **109** 94

Sugano T, Hirai K, Kuroiwa K and Hoh K 1968 *Japan. J. Appl. Phys.* **7** 122

Svensson C 1971 *Proc. IEEE* **59** 1134

Svensson C and Lundström I 1970 *Electron. Lett.* **6** 645

Sze S M 1967 *J. Appl. Phys.* **38** 2951

Szydlo N and Poirier R 1971 *J. Appl. Phys.* **42** 4880

Taft E A 1971 *J. Electrochem. Soc.* **118** 1341

Tarui Y, Komiya Y and Sakamoto T 1973 *Proc. 4th Conf. Solid State Devices, Tokyo 1972, J. Japan. Soc. Appl. Phys.* suppl **42** 145

Thomas J H III and Feigl F J 1972 *J. Phys. Chem. Solids* **33** 2197

Tombs N C, Sewell F A Jr, Comer J J 1969 *J. Electrochem. Soc.* **116** 862

Tombs N C, Wegener H A R, Newman R, Kenney B T and Coppola A J 1966 *Proc. IEEE* **54** 88

Tsujide T 1972 *Japan. J. Appl. Phys.* **11** 62

Tsujide T and Iida K 1972a *Japan. J. Appl. Phys.* **11** 600

—— 1972b *Japan. J. Appl. Phys.* **11** 1599

Tsujide T, Nakaruma S and Ikushima Y 1970 *J. Electrochem. Soc.* **117** 703

Tung S K and Caffrey R E 1965 *Trans. Met. Soc. AIME* **233** 572

—— 1969 *Thin Film Dielectrics* ed F Vratny (New York: Electrochem. Soc.) p 286

Uchida Y, Nishi Y, Nojima I, Tanaka K, Endo N and Tamaru K 1973 *Proc. 4th Conf. Solid State Devices, Tokyo 1972, J. Japan. Soc. Appl. Phys.* suppl **42** 151

Vanstone G F 1972 *Electron. Lett.* **8** 13

Walden R H 1972 *J. Appl. Phys.* **43** 1178

Walden R H and Strain R J 1970 *8th Ann. Proc. Reliability Physics, Las Vegas 1970* ed J B Morris (New York: IEEE) p 23

Wallmark J T and Scott J H 1969 *RCA Rev.* **30** 335

Wegener H A R and Sewell F A Jr 1969 *Tech. Rep.* AFAL-TR-69-187 (Wright–Patterson Air Force Base: Air Force Avionics Lab.)

White M H and Cricchi J R 1972 *IEEE Trans. Electron Dev.* **ED-19** 1280

Williams R 1965 *Phys. Rev.* **140** A569

Woods M H and Tuska J W 1972 *10th Ann. Proc. Reliability Physics, Las Vegas 1972* (New York: IEEE) p 120

Yun B H 1973 *Appl. Phys. Lett.* **23** 152

Yun B H and Arnett P C 1972 *Paper presented at Int. Electron Devices Meeting, Washington 1972*

Zerbst M 1969 *Festkörperprobleme IX—Adv. Solid St. Phys.* ed O Madelung (London/Braunschweig: Pergamon/Vieweg) p 300

Charge-coupled imaging: state of the art

James E Carnes

RCA Laboratories, Princeton, New Jersey 08540, USA

Abstract

This paper is a review of the present state of the art in charge-coupled imaging. After a brief discussion of the various problems associated with conventional X–Y addressing systems for all solid-state, self-scanned imagers, the basic charge-coupled device operation along with speed and efficiency limitations are discussed. Various approaches for charge-coupled line scanners and area imagers are reviewed. Expected performance of charge-coupled imagers is discussed including modulation degradation due to transfer inefficiency and sampling, noise behaviour and sensitivity, interlacing in CCIS, and blooming control. Finally, several experimental devices which have been built to date are discussed and pictures taken with these devices are shown.

1. Background

The apparent advantages of an all solid-state, self-scanned (as opposed to electron beam scanned) image sensor, or television pick-up device, have motivated a large amount of work toward this end over the past ten to fifteen years (Horton *et al* 1964, Weimer *et al* 1969, Weckler 1965, Farnsworth *et al* 1972). These advantages include smaller size, lower power, all solid-state ruggedness, longer lifetime, lower operating voltage—in general, freedom from the problems associated with a vacuum tube and electron beam device.

Prior to the advent of charge transfer devices, all self-scanned imagers incorporated some type of $X–Y$ addressing scheme in which vertical and horizontal conductor lines (either metal or diffusions) were used to address an array of some type of photosensitive elements such as photoconductors or photodiodes. An element was addressed when pulses applied to the horizontal lines and vertical lines coincided at that location.

However, the $X–Y$ addressing scheme faces certain fundamental problems which limit its usefulness. The first is the relatively large capacitance associated with the read-out lines compared with elemental light-sensitive storage capacitance. Since the bus line capacitance increases directly as the number of elements increases, the larger the $X–Y$ addressed system, the more severely the sensitivity of the device is limited by high output capacitance. When transistors are incorporated at each element in an attempt to overcome this problem, a new more serious one of gain variation is encountered. Thus $X–Y$ addressed systems with enough sensitivity and dynamic range for most applications are necessarily limited in the number of resolution elements possible.

Another inherent limitation is the coupling of clocks to the output at each element which cannot be cancelled because of variations in the amount of coupling at each element. Also, small variations in the address pulses applied to the vertical lines are mixed with the signal and generally result in pictures with striations.

Large arrays using $X–Y$ addressing have been built at the research level including a 400×500 bipolar device (Farnsworth *et al* 1972). A 500×1 line scanner and a 50×50 area array are presently available commercially (Weckler 1973).

In 1969 the charge transfer concept for storing and moving charges utilizing silicon integrated circuit technology was introduced by Sangster and Teer (1969): the bucket-brigade shift register. This was followed in 1970 by the announcement of Boyle and Smith (1970) of a new type of charge transfer device utilizing MOS structures: the charge-coupled device (CCD). In this device minority carriers representing the signal are maintained in potential wells formed in silicon by an MOS capacitor. The charge can be stored in the well or transferred from one well to another as the voltage applied to an adjacent metal gate is made more attractive than that where the charge originally was stored. Since the gate voltages tend to repel majority carriers, no recombination occurs and minority carriers can be shifted long distances without undue degradation or loss.

This charge transfer concept has made possible a new scheme for constructing self-scanned imagers. Rather than sequentially addressing each photosensitive element by a switch, the signal charge is shifted in series along a line of MOS capacitors which form a CCD register from the point of photogeneration to a single output amplifier. The charge is shifted from one low-capacitance element to another in sequence and therefore the sensitivity is not limited by high output capacitance regardless of the number of elements. In addition, clock feedthrough occurs only at the output stage at a frequency of twice the video pass band which is easily filtered or otherwise removed. Thus the basic problems of $X-Y$ addressing are not present in a charge-coupled imager (CCI). The CCI has some of its own problems, however, the main one being that the efficiency of the transfer of charge from one element to the next is not always 100%. This tends to degrade modulated signals and can limit the resolution of the device. In addition, close spacing between adjacent MOS capacitor plates must be achieved so that the charge easily spills from one well to the next. This imposes special processing tolerances and/or structures which are not presently made using 'standard' technology.

However, rapid progress has been made in developing charge-coupled imaging in their rather short lifetime and they are already beginning to move into the market place (Amelio 1973).

2. The charge-coupled device (CCD)

2.1. MOS capacitor

Since a CCD is physically just a linear array of closely spaced MOS (metal–oxide–semiconductor) capacitors, it is important to understand the MOS capacitor and how the surface potential V_s (the potential at the Si–SiO$_2$ interface relative to the potential in the bulk of the silicon) depends upon the various parameters involved.

Figure 1 shows a cross-sectional view of an MOS capacitor with a p-type silicon substrate. When a positive step voltage is applied to the gate of such a structure the majority carriers, holes, are repelled and respond within the dielectric relaxation time. This results in a depletion region of negatively charged acceptor states near the surface of the silicon. The applied gate voltage is dropped across the series combination of the oxide and the depletion region in the silicon. Solution of the one-dimensional Poisson's equation subject to the appropriate boundary conditions, and including two-dimensional

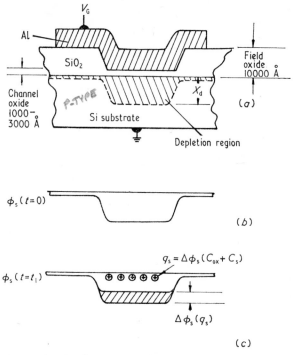

Figure 1. (*a*) Cross-sectional view of MOS (metal–oxide–semiconductor) capacitor. (*b*) Electrical potential at Si–SiO$_2$ interface just after a negative voltage is applied to the gate. (*c*) As positive charge accumulates the depth of the potential well is reduced.

sheets of charge at the Si–SiO$_2$ interface due to fixed oxide charge (Q_{ss}) and signal charge represented by minority carriers (Q_{sig}), shows that the surface potential V_s is given by

$$V_s = V'_G - B\left[\left(1 + \frac{2V'_G}{B}\right)^{1/2} - 1\right] \tag{1}$$

where

$$V'_G = V_G + \left(\frac{X_{ox}}{\epsilon_{ox}}\right)(Q_{ss} + Q_{sig})$$

and

$$B = \frac{qN_A\epsilon_s X_{ox}^2}{\epsilon_{ox}^2} = 0.15\frac{N_A}{10^{15}}\left(\frac{X_{ox}}{1000\ \text{Å}}\right)^2.$$

V_G is the applied gate voltage, Q_{ss} is the fixed oxide charge, Q_{sig} is the signal charge of minority carriers (inversion layer charge), q is the electronic charge in coulombs, N_A is the doping density in acceptors/cm^3, ϵ_s is the dielectric constant of silicon, ϵ_{ox} is the dielectric constant of the oxide layer, and X_{ox} is the thickness of the oxide layer. Equation 1 is a most important one in CCD design.

Just after the step voltage is applied to the gate and in the absence of signal charge Q_{sig} the silicon conduction band at the surface is well below the equilibrium Fermi level and electrons, the minority carriers, will tend to gather there. However, it takes a rather long period of time for thermally generated minority carriers to accumulate in sufficient numbers to return the system to thermal equilibrium. We have measured thermal relaxation times for MOS capacitors ranging from 1 to 100 seconds in good agreement with the predicted values assuming bulk thermal generation of minority carriers. When minority carriers do accumulate at the surface, they start to create an inversion layer which resides within 100 Å of the interface. This negative charge tends to reduce V_s according to equation 1. When V_s goes to zero, no more charge can be accumulated or stored in the potential well. The capacitance of the potential well, C_{well}, consists of two capacitances in parallel: the field oxide capacitance $C_{ox} = \epsilon_{ox}/X_{ox}$, and the capacitance associated with the depletion layer of the silicon, $C_d = (2qN_A\epsilon_s/V_s)^{1/2}$.

Thus the following fluid model of the MOS capacitor emerges: a potential well for minority carriers can be created by applying a step voltage to the gate and this well will take a relatively long period of time to accumulate charge thermally. For times much shorter than this thermal relaxation time, a potential well exists at the surface, and the depth of this well can be altered by changing the gate voltage. When minority carriers are introduced as signal charge in the potential well they tend to reduce the depth of the well according to Q_{sig}/C_{well} so they tend to fill up the well much like fluid in a container.

2.2. Basic charge transfer action

A three-phase CCD is just a line of these MOS capacitors spaced very close together with every third one connected to the same gate, or clock voltage, as shown in figure 2(a). If a higher positive voltage is applied to the ϕ_1 clock line than to ϕ_2 and ϕ_3, the surface potential variation along the interface will be similar to figure 2(b). If the device is illuminated by light, charge will accumulate in these wells. Charge can also be introduced electrically at one end of the line of capacitors from a source diffusion controlled by an input gate. To transfer this charge to the right to the position under the ϕ_2 electrodes, a positive voltage is applied to the ϕ_2 line. The potential well there initially goes deeper than that under a ϕ_1 electrode, which is storing charge, and the charge tends to move over under the ϕ_2 electrodes. Clearly, the capacitors have to be close enough so that the depletion layers overlap strongly and the surface potential in the gap region is a smooth transition from the one region to the other. Next, the positive voltage on the ϕ_1 line is removed to a small positive DC level, enough to maintain a small depletion region, increasing the surface potential under the ϕ_1 gates in the process. Now the ϕ_2 wells are deeper, and any charge remaining

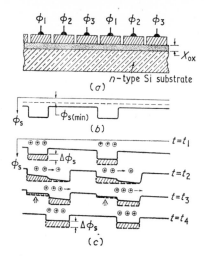

Figure 2. (*a*) Cross-sectional view of a three-phase CCD shift register consisting of closely spaced MOS capacitors. (*b*) Surface potential profile when negative voltage applied to phase 1 gates only. (*c*) Movement of charge as voltage is applied to phase 2 (t_2) and removed from phase 1 (t_3).

under ϕ_1 gates spills into the ϕ_2 wells. The charge, at least most of it, now resides one third of a stage to the right under the ϕ_2 gates. The charge is prevented from moving to the left by the barrier under the ϕ_3 gates. A similar process moves it from ϕ_2 to ϕ_3 and then from ϕ_3 to ϕ_1. After one complete cycle of a given clock voltage the charge pattern moves one stage (three gates) to the right. No significant amount of thermal charge accumulates in a particular well because it is continually being swept out by the charge transfer action.

The charge being transferred is eventually shifted into a reverse-biased drain diffusion and from there it is returned to the substrate. The charging current required once each cycle to maintain the drain diffusion at a fixed potential can be measured to determine the signal magnitude (current sensing) or a re-settable floating diffusion or floating gate which controls the potential of a MOSFET gate can be employed (voltage sensing) (Kosonocky and Carnes 1971b).

2.3. Limitations on speed and efficiency

Clearly, 100% of the charge cannot move instantaneously from one potential well to another. Also, some of the charge gets trapped in fast interface states at each site and cannot move at all. Therefore, in a given clock period, not quite all of the charge is transferred from one well to the next. The fraction of the total that is transferred (per gate) is called the transfer efficiency η. The fraction left behind is the loss per transfer or transfer inefficiency, denoted ϵ, so that $\eta + \epsilon = 1$. Because η determines how many transfers can be made before the signal is seriously distorted and delayed, it is a very important figure of merit for a CCD. The maximum achievable value for η is limited by how fast the free charge can transfer between adjacent gates and how much of the charge gets trapped at every gate location by fast interface states.

Three separate mechanisms cause the free charge to move from one well to another: self-induced drift, thermal diffusion and fringing field drift. Self-induced drift (Engeler *et al* 1970) is a charge repulsion effect which is only important at large signal charge densities ($\geqslant 10^{10}$ charges/cm^2). Thermal diffusion results in an exponential decay of charge under the transferring electrode (Kim and Lenzlinger 1971) with time constant

$$\tau_{th} = \frac{L^2}{2 \cdot 5 D} \tag{2}$$

where L is the centre-to-centre electrode spacing, and D is the diffusion constant. Fringing field drift can help to speed up the charge transfer process considerably. The fringing field is the electric field along the direction of charge propagation at the Si–SiO$_2$ interface. This field will vary with distance along the gate with the minimum occurring at the centre of the transferring gate. The magnitude of the fringing field increases with increasing oxide thickness and gate voltage and decreases with increasing gate length and doping density (Carnes *et al* 1971). The effect of the fringing field upon charge transfer is difficult to assess analytically. A computer simulation of the transfer process under influence of strong

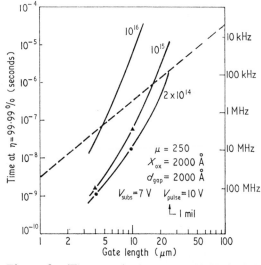

Figure 3. Time required to empty 99·99% of charge against gate length for different substrate doping levels, based upon computer simulation of charge transfer. Broken line is for thermal diffusion only. Solid lines include the effect of fringing field drift. Right ordinate is frequency appropriate for three-phase.

G

fringing fields has indicated that the charge remaining under the transferring electrode still decays exponentially with time. Figure 3 shows the time required to reach $\eta = 99 \cdot 99\%$ as a function of gate length for various substrate doping densities as determined by computer simulation (Carnes *et al* 1972). According to these calculations for a p-channel CCD, $\eta = 99 \cdot 99\%$ is possible at a clock frequency of 10 MHz with gate length $L = 7$ μm and substrate doping of 10^{15} cm^{-3}. This assumes that trapping effects are negligible.

Charges can be lost from the signal into fast interface states because while the filling rate of these states is proportional to the number of free carriers the empty rate depends only upon the energy level of the interface state. Thus, even though a roughly equal amount of time is available for filling as for emptying, many of the interface states can fill much faster than they can empty, and thus release the trapped charge into trailing signal packets (Carnes and Kosonocky 1972a, Tompsett 1973). This type of loss can be minimized by continually propagating a small zero-level charge or fat zero through the device. This tends to keep the states filled so they do not have to be filled by the signal charge. An analytical expression for fractional loss into fast interface states ϵ_s is given by:

$$\epsilon_s = \left(\frac{1/n_{s,0}}{(n_s/n_{s,0}) + 1} \right) kTN_{ss} \ln \left(1 + \frac{2f}{k_1 n_{s,0}} \right) \tag{3}$$

where $n_{s,0}$ is the fat zero carrier density in charges/cm^2, n_s is the signal density in charges/cm^2, kT is in units of eV (0·026 at room temperature), N_{ss} is the fast state density in states/(eV cm^2), f is the clock frequency, and k_1 is a constant depending upon the trapping cross section ($\sim 10^{-2}$ cm^2 s^{-1}).

The effectiveness of fat zero charge in reducing trapping losses in surface channel devices is limited by trapping along the edges since the rounded edges of the well are not exposed to the fat zero charge. Thus the minimum loss per transfer is limited by edge effect trapping and can be approximated according to the following equation (Kosonocky and Carnes 1973b):

$$\epsilon_{edge} = 3 \cdot 9 \times 10^{-4} \left(\frac{1}{W_{mil}} \right) \left(\frac{N_{ss}}{10^{10}} \right) \left(\frac{10^{15}}{N_D} \right)^{1/2} \tag{4}$$

where W_{mil} is the channel width in mils, N_{ss} is the interface state density in (cm^2 eV)$^{-1}$ and N_D is the substrate doping level.

2.4. Buried channel CCDs

The transfer loss due to interface state trapping and the subsequent requirement for fat zero charge can be eliminated if the potential well could be moved away from the Si–SiO$_2$ interface. This is done in the buried channel CCD (Walden *et al* 1972) by including a thin layer of conductivity type opposite to that of the substrate as shown in figure 4(a). When this layer is completely depleted of majority carriers by applying the appropriate potential to the drain diffusions,

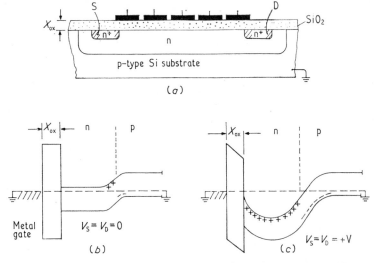

Figure 4. (*a*) Cross-sectional view of buried channel CCD. (*b*) Band diagram in thermal equilibrium. (*c*) When the buried *n*-layer is completely depleted of electrons, a parabolic potential well forms with the minimum located away from the Si–SiO₂ interface.

the depleted layer results in a parabolic potential well as seen in figure 4(*c*). This well can store and transfer charge as described earlier.

Since the signal charge is then no longer subjected to fast interface state trapping, there is no requirement for fat zero. In addition the carrier mobility is higher since the charge transport occurs in the bulk rather than at the surface, and fringing fields are higher since the electrodes are further away. Thus buried channel devices should be faster, not require fat zero and not suffer from edge effect trapping.

On the other hand, buried channel devices involve more complex processing and more critical operation. They also have reduced signal handling capability because the well capacitance is smaller. Bulk trapping states may also affect buried channel operation at certain resonant frequencies.

2.5. *Charge-coupling gate structures*

A wide variety of gating schemes have been introduced for CCDs. Several of the most important are shown in figure 5 and will be discussed in this section.

2.5.1. *Three-phase single-level metal CCDs.*
The three-phase CCD whose operation was described earlier can be made by standard p-MOS or n-MOS techniques (see figure 5(*a*)). Only one level of metal is required in addition to source and drain diffusions. The gap between adjacent electrodes presents the major problems. The gap should be as small as possible to insure good coupling between gates

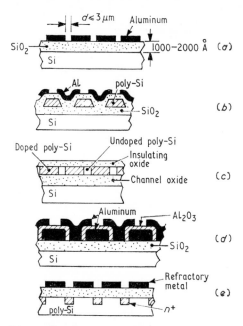

Figure 5. Cross-sectional view of various CCD gate structures: (*a*) Single metal, (*b*) overlapping polysilicon–aluminium, (*c*) doped polysilicon with undoped polysilicon forming the gaps, (*d*) anodized aluminium, and (*e*) conductively connected CCD, the C4D.

without any bumps or barriers in the surface potential. Thus 2 to 3 μm etching between metal electrodes is desirable but is a non-standard process. Furthermore, if the oxide in these gaps is exposed to the ambient air, the surface potential in the gaps is largely uncontrolled, and CCD operation can be adversely affected. Single-level metal devices will, therefore, require some type of overcoat or resistive sea layer over the gap regions to control and stabilize the surface potential there. Another disadvantage of any three-phase system is a topological one. To contact the three-phase electrodes at least one crossover or dig-down structure is needed at each stage.

The main advantage of the three-phase single-metal CCD is that it can be made with the minimum number of processing steps. The price for this is the definition of 2 to 3 μm gaps between the gates.

2.5.2. Two-phase CCDs.   In three-phase CCDs the potential wells are symmetrical and the directionality of charge flow is maintained by the asymmetry of the clocking voltages as seen in figure 2. However, if each potential well had a built-in asymmetry to determine the direction of charge flow, then a symmetrical two-phase clocking scheme could be used to drive the device. The potential well asymmetry required is that the well must be deeper in the direction of charge transfer. This can be achieved by having two thicknesses of oxide under one electrode or by having a variation in substrate doping (see equation 1). A two-phase device has clear topological advantages over a three-phase device since no dig-downs are required for access to the electrodes.

A particularly advantageous method of constructing two-phase CCDs which results in two thicknesses of oxide to provide signal directionality consists of polysilicon gates overlapped by aluminium gates (see figure 5(b)) (Kosonocky and Carnes 1971a, 1973a). After definition, polysilicon gates are partially oxidized to form an insulating layer and to increase the oxide thickness in the region between the gates. The aluminium is deposited and defined so that it overlaps the polysilicon gates as shown. The spacing between electrodes is determined by the polysilicon oxide thickness—typically 2000 Å. No gaps are exposed to ambient and standard alignment (2·5 μm) and etching (5 μm) techniques result in 30 μm per stage (two polysilicon gates and two aluminium gates) spacing.

2.5.3. Other gate structures. Other possible gating structures are shown in figure 5(c–e). The polysilicon structure shown in figure 5(c) can be used for both two- and three-phase devices and avoids exposure of the gaps because of the presence of undoped (insulating) polysilicon in the space between electrodes (Kim and Snow 1972). The electrode region is doped polysilicon. Another method for obtaining overlapping gates consisting of anodized aluminium as the insulating material (Collins *et al* 1972) is shown in figure 5(d). Finally, the conductively connected charge-coupled device (C4D), (Krambeck *et al* 1972) is shown in figure 5(e). The structure has self-aligned ion-implanted n⁺ (for n-channel) regions in the gaps between electrodes. This obviates the need for etching gaps of 3 μm or less.

3. Charge-coupled image sensors

3.1. Line scanners

Two ways by which the optical signal can be introduced into the CCI are illustrated in figure 6. The optical input can be introduced from the top of the substrate through the spaces between non-transparent metal gates (figure 6(a)). Top illumination of the CCI is also possible by transmitting the optical input through transparent gates such as thin polysilicon. An alternative approach (figure 6(b)) consists of thinning the substrate in the optically sensitive area and applying the optical input from the back side of the substrate. Figure 6 also illustrates two ways by which the output can be removed from the CCI array. Output 1 is the current output derived from the drain diffusion D. The low impedance output 2 samples the voltage of the floating diffusion F and is proportional to the charge signal. A similar arrangement employing a floating gate can also be utilized to sense the charge signal (Kosonocky and Carnes 1971b). Let us assume now that an optical input is applied to such a CCD register while

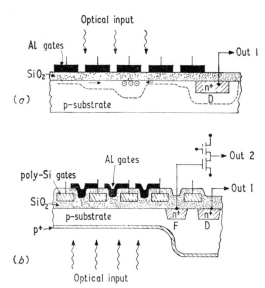

Figure 6. Cross-sectional views of two types of CCD line imagers: (*a*) top illuminated with three-phase single-metal gates and current sensing output; (*b*) thinned back-illuminated, two-phase device with a voltage sensing (floating diffusion) output.

the clock voltages are adjusted so that one potential well is created at each stage along the CCD channel. As suggested in figure 6(*a*), the photogenerated charge will collect in these wells during the optical integration time. At the end of the integration time, the accumulated charge packets representing the integrated optical input are shifted down the CCI register and detected by a single output amplifier. To prevent smearing of the image, the optical integration time should be much larger than the total time required to transfer the detected image from the CCD line sensor. Since all charge elements are amplified by the same amplifier, nonuniformities—usually a problem in optical arrays in which each sensor element uses a separate amplifier—are avoided. Since there is no direct coupling of the clock voltages to the charge signal in the CCD channel, the clock pick-up is limited only to a single output stage. In addition, since only the clock frequency, which is outside of the video bandpass, is used in CCD transfer, clock pick-up is not the problem as it is in X–Y scanned arrays. This type of device in which the CCD register serves to store the photogenerated charge during integration and shift it to the output register during read-out is called an illuminated register device.

A more effective CCD line sensor (or CCI line array) is shown schematically in figure 7(*a*). Here, the optical input can be continuously integrated by the linear array of photosensors. During operation, the detected line image is periodically transferred in parallel to the CCD register from where it is read out serially.

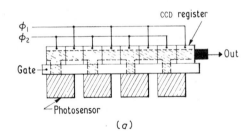

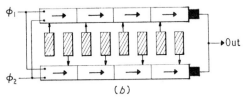

Figure 7. Line arrays with separate photosensors: (*a*) single CCD read-out register; (*b*) two registers reading out alternate photoelements.

It is essentially an analogue parallel-to-series converter, with time-integrated optical input and electrical output. A dual CCD channel line sensor is shown in figure 7(*b*). The operation of this CCI is the same as for the line sensor described above except that the detected image is transferred into two parallel CCD registers and then combined into a single output line. The advantage of the dual CCD channel line sensor over the line sensor shown in figure 7(*a*) is double resolution and fewer CCD transfers.

3.2. Area arrays

One way of implementing an area CCI is illustrated in figure 8. This array can be visualized as a parallel array of the previously described linear arrays whose outputs are transferred in parallel into a single output register. The operation of this CCI is as follows. Once every frame time the charge signal detected by the photosensor array is transferred into the vertical CCD channels, which are not photosensitive. Then, the entire detected image is shifted down in unison by clock A and transferred into the output register one (horizontal) line at a time. The horizontal lines are then transferred out from the output register by the high frequency clock B before the next horizontal line is shifted in. This approach to area imaging is called the interline system.

Another frame transfer CCI that does not require separate photosensors is shown in figure 9 (Tompsett *et al* 1971). In this CCI, the photosensor function is performed by an additional photosensitive CCD array. This system is composed of three functional parts: the photosensitive array, a temporary storage array and the output register. The optical image is detected by the photosensitive array. Then, assuming a TV format with $\frac{1}{60}$s frame time, the detected image is trans-

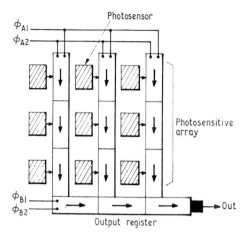

Figure 8. Interline transfer area imaging system.

ferred into the temporary storage array by clocks A and B, during the vertical blanking time (900 μs). From there it is shifted down one horizontal line at a time into the output register and transferred out by the high speed clock C. The time available for parallel loading of the output register corresponds to the horizontal line retrace time of 10 μs, which leaves 50 μs for the read out of the horizontal line from the output register.

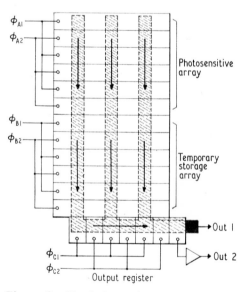

Figure 9. Vertical transfer system with separate store.

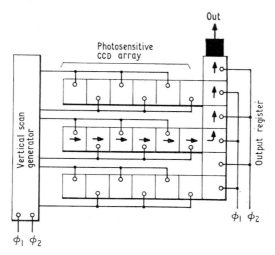

Figure 10.   Horizontal line-by-line area imaging system.

A third type of CCD area array (figure 10) was made originally at RCA Laboratories, using bucket-brigade shift registers (Kovac *et al* 1971). This system consists of a parallel array of photosensitive, horizontal, CCD channels, all leading into a single output register. This CCI operates by transferring, under the control of the vertical clock generator, one horizontal line at a time from the photosensitive CCD array to the output register and out.

4. Performance limitations of CCIs

4.1. MTF

The main limitation of charge-coupled imagers is the degradation of signal modulation which can occur after the large number of transfers required. The severity of this problem depends upon the overall loss $N\epsilon$, where N is the total number of transfers and ϵ is the inefficiency per transfer. If a light pattern with spatial frequency K is incident upon a CCI, by the time the signal reaches the output the modulation of the pattern is degraded by a factor given (Berglund 1971) by the modulation transfer function (MTF):

$$\mathrm{MTF}(K) = \exp\left[-\epsilon N\left(1 - \cos\frac{\pi K}{2K_{\max}}\right)\right]. \tag{5}$$

$K_{\max}$ is the maximum spatial frequency appropriate for the centre-to-centre spacing between elements, L, where

$$K_{\max} = \frac{\pi}{L}. \tag{6}$$

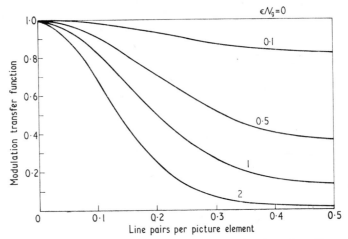

Figure 11.   Modulation transfer function (MTF) against line pairs per picture element due to transfer loss.   The parameter is ϵN_g, the total fractional loss after N_g transfers.

Figure 11 shows a plot of MTF due to transfer inefficiency against line pairs per picture element with ϵN as a parameter.   Since the overall system MTF is the product of all the individual MTFs, the maximum permissible value for ϵN will depend upon other components of the system. ϵN values of unity should be satisfactory in most cases.

Clearly, since some elements require more transfers to reach the output the resolution of a large CCI may degrade as a function of distance from the output. This should not be a discernible effect if ϵ is low, however.   Also since in general the number of horizontal transfers required for a given element is different from the number of vertical transfers, the MTF may be different for the two directions.

Another source of signal modulation degradation in a CCI is the sampling process which occurs when the modulated light intensity input is detected by sampling the intensity at discrete points.   For a top-illuminated single-metal CCI, the light intensity is sampled only in the gaps between electrodes.   Sequin (1973) shows MTF curves for several different sampling cases.   The two most frequently encountered are given approximately by

$$\text{MTF}(K)_{\text{sampling}} = \frac{\sin (\pi K/2 K_{\text{max}})}{\pi K/2 K_{\text{max}}}. \tag{7}$$

This means that the MTF due to sampling is down to approximately 0·6 at $K = K_{\text{max}}$.

4.2. Noise and sensitivity

The CCD is inherently a low noise device (Barbe 1972, Carnes and Kosonocky

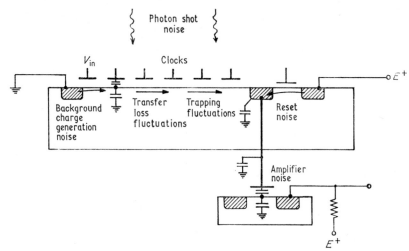

Figure 12. Charge-coupled device noise sources.

1972b) and therefore has the capability of providing very sensitive imager performance (Carnes and Kosonocky 1972c). The various noise sources are shown in figure 12. Its main advantage over the silicon vidicon is the much lower output capacitance of the CCI, which typically is on the order of or less than one pF. This reduces the effect of thermal noise at the output which limits silicon vidicon performance.

The transfer of charge from one well to another in a CCD is inherently a low noise process if the charge is completely emptied (except for the small fraction ϵ left behind).

In surface channel CCDs, the interface states are also a source of noise because while they may be filling and emptying to the same level each period (and therefore not causing any signal loss) there are statistical fluctuations which add in quadrature as the charge is transferred.

The noise associated with the transfer process has a special property in that the gain or loss of charge of a given packet due to fluctuations must be compensated for by a loss or gain in charge in the two adjacent packets. Thus the output noise signal is not completely uncorrelated but has some correlation at plus or minus one period of the clock. Thus, as shown by Thornber and Tompsett (1973), the spectral power density of transfer noise is not 'white' as for uncorrelated noise sources, but has a frequency dependence given by

$$P(f) \propto \left(1 - \cos \frac{2\pi f}{f_c}\right). \tag{8}$$

This function is shown in figure 13. Most of the transfer noise power is concentrated at higher frequencies. Experimental measurement of CCD noise has confirmed this type of transfer noise spectrum (Carnes *et al* 1973).

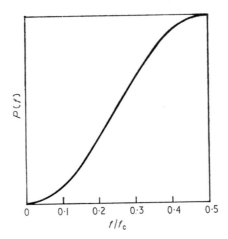

Figure 13. Spectral density of transfer noise in a CCD.

Calculations and measurements indicate that the sensitivity of surface channel CCIs will be limited by fast interface state trapping noise and shot noise of background charge. Carnes and Kosonocky (1972c) have shown that the fluctuation in carriers per well for typical dimensions and 2000 transfers (appropriate for a 500×500 element device) should be about 1000 electrons. This limits the sensitivity to about 1000 electrons and, since full well signals of 10^6 are achievable, limits the dynamic range to about 1000.

Calculations assuming a completely uniform 500×500 CCI with no spatial variations in average dark current and excluding MTF effects indicate that surface channel CCIs have the fundamental capability of achieving very sensitive performance—somewhat better than the silicon vidicon.

Since buried channel devices have no interface state trapping noise and no need for fat zero, they should be capable of more sensitive CCI operation than surface channel devices.

The sensitivity of actual area devices at room temperature has been limited to date by spatial nonuniformities in dark current which tend to overwhelm small video signals. This problem applies to both surface and buried channel. Device cooling improves this problem as should improved processing.

4.3. Interlacing

To be compatible with standard TV scanning, an imager must be capable of interlace—the reading of every other horizontal line in successive field times ($\frac{1}{60}$s for US system, $\frac{1}{50}$s for PAL). This system was introduced originally to eliminate the flicker the eye could detect when the picture was refreshed at $\frac{1}{30}$s time intervals. In a charge-coupled imager, in addition to reducing flicker, interlacing actually increases the resolution of the device (Sequin 1973).

Consider a three-phase vertical transfer with separate store (figure 9) CCI. While three gates are required to shift one charge packet, the charge can be stored under any one gate during the integration time. Thus the built-in structure for storage is three times that for transfer. Interlacing in a three-phase CCI is illustrated in figure 14. During field 1 integration, phase 1 electrodes are maintained at a higher potential than phase 2 or 3. Thus the centre of charge collection is under phase 1 gates. Note, however, that the charge generated under phase 2 and 3 gates also is collected under phase 1 since phase 2 and 3 gates are biased into depletion so that recombination is low and photogenerated electrons will diffuse to the phase 1 wells. All light-generated carriers are collected. During field 2, phase 2 and phase 3 are kept higher so the centre of

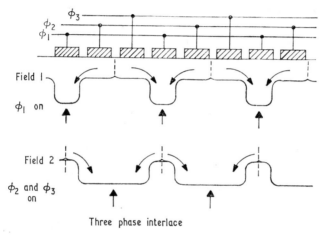

Three phase interlace

Figure 14. Cross-sectional view of a three-phase vertical transfer imager indicating how interlacing can be accomplished. Vertical arrows indicate the centre of gravity of collected charge.

charge collection is moved by one half of a cell as indicated by the arrows. Thus in successive fields the light intensity is sampled at centres shifted by one-half of a resolution cell. This effectively doubles the sampling spatial frequency from $2\pi/L$ to $4\pi/L$ and correspondingly reduces the distortion introduced by fold-over effects (Sequin 1973). The net result is about double resolution. Thus a three-phase vertical transfer system with N three-gate resolution elements in the imaging region in the vertical direction is capable of almost $2N$ TV lines per picture height. However, in the horizontal direction one must have one channel for each resolution element. In the vertical transfer system, interlacing results in increased resolution with all light collected and an integration time for each field of $\frac{1}{60}$ s.

Interlace is also possible in the interline transfer system (figure 8) but since the separate photosensor elements are fixed, $2N$ sensors must be present on the chip to achieve $2N$ resolvable elements. Every other photosensor is read out in successive fields. The integration time for one field is then $\frac{1}{30}$s.

4.4. Blooming control

Since CCDs are designed to permit motion of minority carriers over long distances without recombination, it is expected that excess charge due to point light overloads will spread down the CCD channels and saturate large regions of the device. This effect is known as blooming and also is observed in silicon vidicons. Several techniques have been devised and successfully operated in line arrays for controlling blooming in CCDs (Kosonocky *et al* 1972).

The basic idea in blooming control structures is to provide diffused charge drains to remove the excess charge. The drains, or blooming busses, are parallel to the CCD channel and separated from it by a potential barrier, called the blooming barrier, which is designed to allow excess charge to flow from the channel into the drain before it can spread down the channel. Figure 15 shows

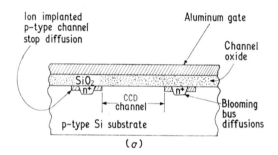

(a)

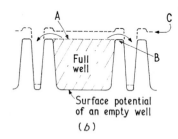

(b)

Figure 15. (a) Cross-sectional view of CCD with blooming control consisting of n^+ drains and ion-implanted p-type barriers. (b) Potential profile indicating how excess charge will flow into the drains before spreading down the channel. A denotes the surface potential of a full well. B indicates the potential of the blooming barrier while C is the surface potential of one of the transfer gates. Charge will spill over B into the drain bus before spilling down the register.

a cross section across a channel in which the barrier between channel and n+ blooming bus diffusions is achieved by a p-type region introduced by ion implantation. Part (*b*) of the figure shows how excess charge spills into the drains before it can overcome the barriers of the CCD gates which are 'OFF' (see dotted lines). Other means of establishing the barrier are possible, including a thick oxide and a polysilicon field shield. No known area array devices have yet been built which incorporate blooming control structures.

5. The present status of charge-coupled imaging

In this section the most recent charge-coupled imaging devices which have been publicly announced will be reviewed.

5.1. Line arrays

The announcement of the first production charge-coupled device was made at the IEEE International Convention in March 1973 (Amelio 1973) exactly three years after the introduction of the CCD at that same conference (Boyle and Smith 1970). The device, built by Fairchild Camera and Instrument Corporation, is a 500 photoelement line scanner. The 500 photoelements are read out alternately by two 250-stage, three-phase, buried channel devices with polysilicon gates with undoped gaps. Figure 16 shows four line scan pictures (mechanical movement in vertical direction) made with this device. The pictures were made by decreasing the light intensity by factors of 10 so that the light intensity of the lower right picture is only 1000th of that of the top left. Subsequent measurements showed a dynamic range of 3000. The salt and pepper noise in the low intensity picture was introduced by the on-chip amplifier

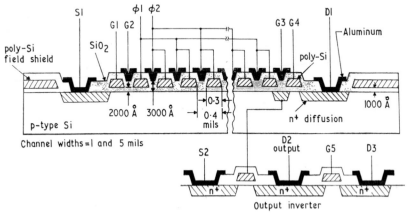

Figure 18. Cross-sectional view of the RCA Laboratories' 500 × 1, two-phase line array.

(G F Amelio private communication). The low light regions in this picture resulted in only 150 electrons per charge packet. This sequence of photos clearly demonstrates the low noise, high sensitivity and high dynamic range capabilities of the buried channel charge-coupled device.

Workers at RCA Laboratories have built a 500-stage illuminated register line array using both three-phase, single-level metal and two-phase polysilicon–aluminium technologies. A line scan picture taken with the three-phase device is shown in figure 17. The gate lengths are 7·5 μm with 2·5 μm gaps for a resolution element centre-to-centre spacing of 30 μm. A cross-sectional drawing of the two-phase 500×1 chip is shown in figure 18. The gates are 7·5 μm long with 2·5 μm gaps for a centre-to-centre spacing of 20 μm. A line scan picture (mechanical motion in the vertical direction) is shown in figure 19.

The largest CCD line array to date was made at Bell Telephone Laboratories: a 1500 photoelement device read out by two four-phase 750-stage surface channel CCD registers (Sealer *et al* 1973). The chip is shown in figure 20 and the resultant image of a typed page made with this device is shown in figure 21.

5.2. Area arrays

Development of area arrays has also been rapid. While the first area array charge transfer imager, a 32×44 horizontal line device using bucket-brigade registers was built by RCA Laboratories (Kovac *et al* 1971), Bell Telephone Laboratories announced the first CCD area array—a 64×106 vertical transfer system with separate store having an overall size of 106×128 (Sequin *et al* 1973). Three of these devices have recently been utilized to build a colour CCD camera (M F Tompsett and E J Zimany, private communication). The light is split by a prism and the different colour components imaged on the three devices as seen in figure 22. The image is shown in figure 23. (This figure is not in colour. The colour version can be seen on the cover of the January 18, 1973 issue of *Electronics*.)

As with line arrays, Fairchild has been the first to announce a commercially available area array product—a 100×100 buried channel array using interline transfer (Amelio 1973). This device was announced as a commercial product in the August 30, 1973 issue of *Electronics*. It employs a two-phase gating system. A picture taken with this device is shown in figure 24. This array also displays a dynamic range of 1000 to 1 (G F Amelio private communication).

The largest area imager announced to date is a 128×160 three-phase vertical transfer device built by RCA (Kovac *et al* 1973). Including separate store this device is 256×160 which involves 122880 separate MOS capacitors and has the capability of storing 40960 bits of information. The chip size is 0·79 cm by 0·95 cm and is shown in figure 25. The output stage is magnified in figure 26. A picture taken with this device and displayed on an unaltered standard (USA) TV display is shown in figure 27.

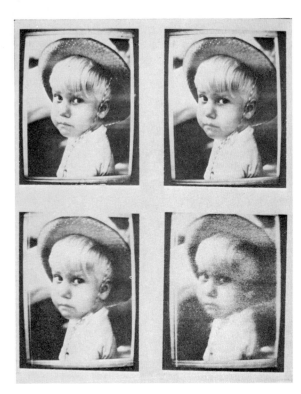

Figure 16. Four line scan pictures made with the Fairchild 500×1 device (mechanical motion in the vertical direction). Upper left picture is taken at saturating (full well) light intensity. The light intensity is reduced by a factor of 10 for each of the other pictures so that the light intensity for the bottom right picture is 1000 times lower than for the top right picture.

Figure 17. Line scan picture made with the 500×1, three-phase illuminated register device built at RCA Laboratories.

Figure 19. Picture taken with 500×1, two-phase CCD built by RCA Laboratories.

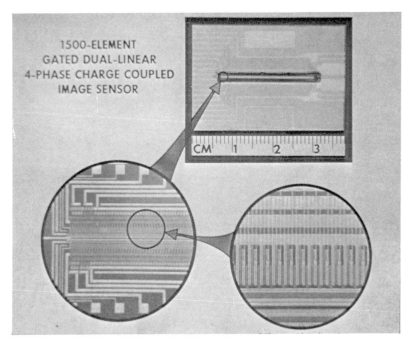

Figure 20. Several views of the 1500 photoelement line imager chip built by Bell Telephone Laboratories.

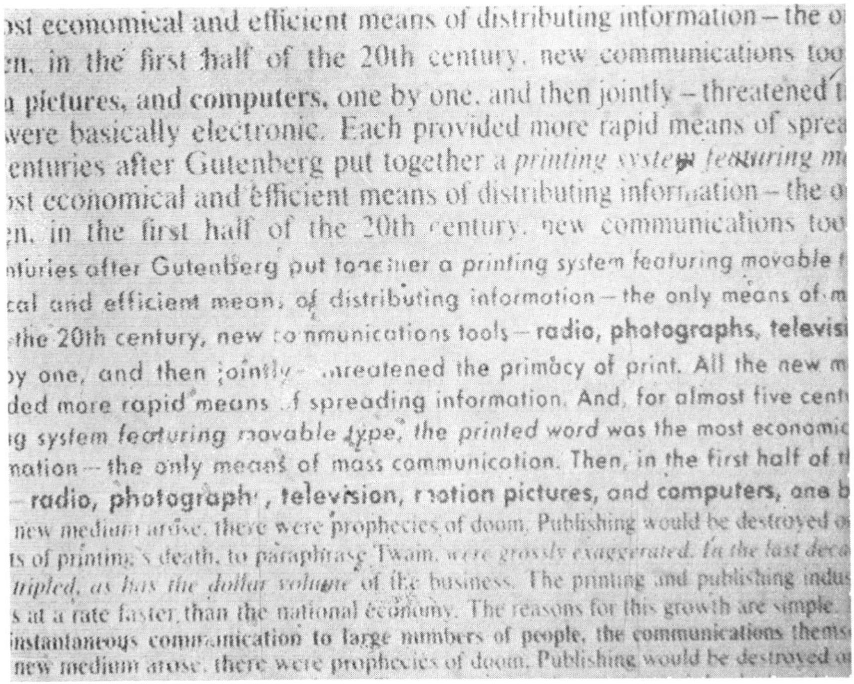

ıst economical and efficient means of distributing information – the o
en, in the first half of the 20th century, new communications too
a pictures, and computers, one by one, and then jointly – threatened t
were basically electronic. Each provided more rapid means of sprea
enturies after Gutenberg put together a *printing system featuring m*
ıst economical and efficient means of distributing information – the o
en, in the first half of the 20th century, new communications too
nturies after Gutenberg put together a printing system featuring movable t
cal and efficient means, of distributing information – the only means of m
the 20th century, new communications tools – radio, photographs, televisi
by one, and then jointly threatened the primacy of print. All the new m
ded more rapid means of spreading information. And, for almost five cent
ıg system featuring movable type, the printed word was the most economic
nation – the only means of mass communication. Then, in the first half of t
– radio, photograph, television, motion pictures, and computers, one b
new medium arose, there were prophecies of doom. Publishing would be destroyed o
ıs of printing's death, to paraphrase Twain, were grossly exaggerated. In the last deca
tripled, as has the dollar volume of the business. The printing and publishing indus
s at a rate faster than the national economy. The reasons for this growth are simple.
instantaneous communication to large numbers of people, the communications thems
new medium arose, there were prophecies of doom. Publishing would be destroyed o

Figure 21. Portion of normal printed page scanned by the 1500×1 photoelement Bell Telephone Laboratories device. Magnification is one-to-one.

Figure 22. Photograph of the first CCD colour camera. It was built at Bell Telephone Laboratories.

Figure 23. Black and white version of Bell Telephone Laboratories colour image. Full colour version can be found on the cover of the January 18, 1973 issue of *Electronics*. (Copyright McGraw-Hill Inc, 1973).

Figure 24. Image taken with Fairchild's 100×100 area array utilizing an interline transfer system and buried channel registers. Dotted picture resulted because clock feedthrough was not filtered.

Figure 25. Photograph of the RCA 128×160 three-phase, surface channel imager using the vertical transfer system. This is the largest charge-coupled imager announced to date. Chip dimensions are 0.7×0.95 cm.

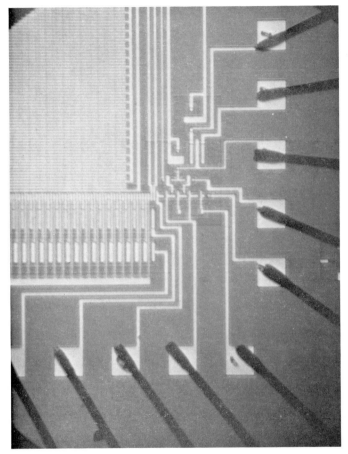

Figure 26. An expanded view of the output stage of the RCA 128×160 imager.

Figure 27. The author as viewed by the RCA 128×160 imager and displayed on a standard unaltered TV monitor. Moiré patterns in the fan indicate that full geometric resolution is being achieved.

6. Conclusions

The charge-coupling concept presents an alternative technique for achieving all solid-state, self-scanned imaging. It is free from the clock feedthrough, high output capacitance, and uniformity problems of $X-Y$ addressed systems. The CCI is inherently a low noise device with low output capacitance and appears capable of more sensitive operation than the silicon vidicon. A cooled buried channel device may exceed I-SIT performance.

The main limitation of the charge-coupled approach is the large number of transfers required which demands high transfer efficiency to preserve signal modulation. However, transfer inefficiencies of 10^{-4} have been achieved and thus devices requiring 1000 transfers should be possible without discernible signal degradation.

The ultimate factor affecting the eventual success of charge-coupled imagers in the market place depends upon the ability of manufacturers to fabricate the large chips required (approaching 2 cm on a side) with yields high enough to permit low cost. Large $X-Y$ addressed devices (400×500) (Farnsworth *et al* 1972) were built in the laboratory, but on a production basis the size of $X-Y$ devices presently is limited by current LSI technology.

Perhaps the inherent performance advantages of charge-coupled imagers will motivate the industry to develop the large chip technology required. If so, CCIs should become quite prevalent in closed circuit and medium quality TV systems within 2–4 years and eventually may be used in broadcast quality systems.

Acknowledgments

Some of the work at RCA Laboratories on three-phase imagers was supported by the US Naval Electronics Systems Command while work on the two-phase device was supported by the National Aeronautics and Space Administration. The author is also indebted to many persons who helped materially in the preparation of this paper including M F Tompsett of Bell Telephone Laboratories and G F Amelio of Fairchild Camera and Instrument Corporation who kindly supplied pictures by and of their respective devices. Many persons within RCA also provided invaluable assistance including P A Levine and R L Rogers who provided pictures, and M G Kovac, W F Kosonocky and K H Zaininger, all of whom provided valuable assistance with manuscript writing.

References

Amelio G F 1973 *Physics and applications of charge-coupled devices, paper presented at IEEE Intercon, New York, March, 1973*
Barbe D F 1972 *Noise and distortion considerations in charge-coupled devices, Electron. Lett.* **8** 207

H

Berglund C N 1971 *Analog performance limitations of charge-transfer dynamic shift registers, IEEE J. Solid St. Circuits* **SC-6** 391

Boyle W S and Smith G E 1970 *Charge-coupled semiconductor devices, Bell Syst. Tech. J. Briefs* **49** 587

Carnes J E and Kosonocky W F 1972a *Fast interface state losses in charge-coupled devices, Appl. Phys. Lett.* **20** 261

—— 1972b *Noise sources in charge-coupled devices, RCA Rev.* **33** 327

—— 1972c *Sensitivity and resolution of charge-coupled imagers at low light levels, RCA Rev.* **33** 607

Carnes J E, Kosonocky W F and Levine P A 1973 *Measurement of noise in charge-coupled devices, RCA Rev.* **34** (December) to be published

Carnes J E, Kosonocky W F and Ramberg E G 1971 *Drift-aiding fringing fields in charge-coupled devices, IEEE J. Solid St. Circuits* **SC-6** 322

—— 1972 *Free charge transfer in charge-coupled devices, IEEE Trans. Electron Dev.* **ED-19** 798

Collins D R, Shortes S R, McMahon W R, Penn T C and Bracken R C 1972 *Double level anodized aluminum CCD, paper presented at IEEE Int. Devices Meeting, Washington, DC, December 1972*

Engeler W E, Tiemann J J and Baertsch R D 1970 *Surface charge transport in silicon, Appl. Phys. Lett.* **17** 469

Farnsworth D L, Irwin E L and Huggins C T 1972 *TV resolution solid-state array camera, Government Microcircuit Applications Conference, San Diego, October 1972*

Horton J W, Mazza R V and Dym H 1964 *The Scansistor—a solid-state image scanner, Proc. IEEE* **52** 1513

Kim C-K and Lenzlinger M 1971 *Charge transfer in charge-coupled devices, J. Appl. Phys.* **42** 3586

Kim C-K and Snow E H 1972 *P-channel charge-coupled device with resistive gate structure, Appl. Phys. Lett.* **20** 514

Kosonocky W F and Carnes J E 1971a *Charge-coupled digital circuits, Digest Tech. Papers, IEEE Solid-State Circuit Conf., February 1971* p162

—— 1971b *Charge-coupled digital circuits, IEEE J. Solid-St. Circuits* **SC-6** 314

—— 1973a *Two-phase charge-coupled devices with overlapping polysilicon and aluminium gates, RCA Rev.* **34** 164

—— 1973b *Two-phase charge-coupled devices with overlapping polysilicon and aluminum gates, paper presented at Charge-Coupled Device Applications Conference, San Diego, September 1973*

Kosonocky W F, Kovac M G, Weimer P K and Carnes J E 1972 *Blooming control and dynamic range in charge-coupled imagers, paper presented at IEEE Int. Electron Dev. Meeting, Washington, DC, December 1972*

Kovac M G, Shallcross F V, Pike W S and Weimer P K 1971 *Image sensors based upon charge transfer by integrated bucket-brigades, paper presented at IEEE Int. Electron Dev. Meeting, Washington, DC, October 1971*

—— 1973 *Design, fabrication and performance of a 128 × 160 element charge-coupled image sensor, paper presented at Charge-Coupled Device Applications Conference, San Diego, September 1973*

Krambeck R H, Strain R J, Smith G E, and Pickar K A 1972 *A conductively connected charge-coupled device, paper presented at IEEE Int. Electron Dev. Meeting, Washington, DC, December 1972*

Sangster F L J and Teer K 1969 *Bucket-brigade electronics—new possibilities for delay, time-axis conversion and scanning, IEEE J. Solid St. Circuits* **SC-4** 131

Sealer D A *et al* 1973 *IEEE Trans. Electron Dev.* to be published

Sequin C H 1973 *Interlacing in charge-coupled imaging devices, IEEE Trans. Electron Dev.* **ED-20** 535

Sequin C H *et al* 1973 *A charge-coupled area image sensor and frame store, IEEE Trans. Electron Dev.* **ED-20** 244

Thornber K K and Tompsett M F 1973 *Spectral density of noise generated in charge transfer devices, IEEE Trans. Electron Dev.* **ED-20** 456

Tompsett M F 1973 *The quantitative effects of interface states on the performance of charge-coupled devices, IEEE Trans. Electron Dev.* **ED-20** 45

Tompsett M F *et al* 1971 *Charge-coupled imaging devices: experimental results, IEEE Trans. Electron Dev.* **ED-18** 992.

Walden R H, Krambeck R H, Strain R J, McKenna J, Schryer N L and Smith G E 1972 *The buried channel charge-coupled device, Bell Syst. Tech. J.* **51** 1635

Weckler G P 1965 *A silicon photodevice to operate in a photon flux integration mode, paper presented at Int. Electron Dev. Meeting, Washington, DC, October 1965*

—— 1973 *Solid-state image sensing with photodiode arrays, paper presented at IEEE Intercon, New York, March 1973*

Weimer P K, Pike W S, Sadasiv G, Shallcross F W, and Meray-Horvath L 1969 *Multi-element self-scanned mosaic sensors, IEEE Spectrum* **52**

Bipolar devices for low power digital applications—progress through new concepts

Horst H Berger

IBM Deutschland GmbH, Dept 3190, D-703 Boeblingen,
Schoenaicher Str. 220, Federal Republic of Germany

Abstract

The present supply of digital monolithic circuits suggests that bipolar devices are not useful for the low power range ($P \lesssim 300$ μW/circuit). This paper shows in contrast that bipolar devices can be utilized to obtain low power digital circuits with excellent speed/power ratios. The usage of lateral pnp transistors as load devices, and of Schottky barrier diodes and superintegrated devices for low power dissipation at high density is discussed. With the IIL/MTL logic circuit it is shown that for example a 20 ns delay at a power dissipation of only 10 μW is within reach. The high density and low power requirements of approximately 10 kbit memory chips can be achieved with *static* stable bipolar memory cells that offer much better performance than dynamic cells usually required with MOSFETs above 2 kbit/chip.

1. Introduction

'Power in a squeeze' has been the cover statement of a recent *IEEE Spectrum* issue. Although it was meant for the big energy crisis, it underlines our power crisis in digital monolithic circuits as well.

This crisis is either primarily that of too much power consumption or of too much power dissipation. In medical electronics and portable equipment energy sources are small, demanding very low power *consumption*. In large scale integration power *dissipation* per circuit is limited because of the high circuit density and the limited cooling capability per unit of surface area.

If we want to have cheap air convection cooling and if junction temperature is to stay below 100 °C, power dissipation density must be kept below about 100 mW cm^{-2}. With forced air cooling, about 300 mW cm^{-2} can be tolerated. Hence, if we strive for more than 1000 basic circuits/cm^2 of package surface, we have to keep power dissipation per circuit below 300 μW or preferably 100 μW (see figure 1).

The electronic wristwatch provides an example of an extremely low power consumption requirement. Having 20 or more basic circuits it must not

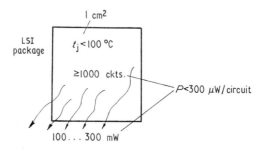

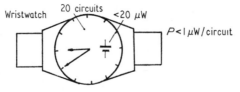

Figure 1. Power dissipation and power consumption limits of digital circuits.

consume more than about 20 μW total.   Thus we get less than 1 μW/circuit here.

With these considerations, a low-power digital circuit can be characterized by a power dissipation of definitely less than 300 μW.

It appears almost superfluous to raise the question at all, which devices—MOS or bipolar ones—are best suited for digital circuits in the low power range, as the broad market has already given a clear answer: 'MOS for low power'.

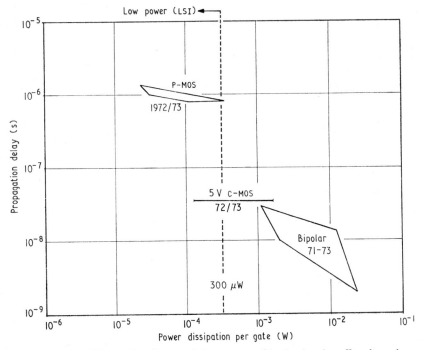

Figure 2.   Power–delay ranges of logic circuits offered on the market.

Figure 2 shows this market situation for the logic circuits.   Average propagation delay is plotted against average power dissipation per gate for various products on the market.   The diagram is certainly not complete, but the main point is that there are no bipolar logic circuits on the market offered in the low power range below 300 μW/gate, not even below 1 mW/gate.   The situation is similar in the monolithic memory market, where MOS devices dominate the low power applications.

One could conclude now with the superficial but plausible sounding explanation that MOS devices are suited for low power because of their high impedance, but bipolars are not, because of their low impedance and their base current requirement.   However, the title of this paper is *Bipolar* devices for low power digital applications and this contrary statement will be substantiated in the following.

First, logic circuits will be discussed and the role of bipolar devices in getting low power there.   The second part of this paper will deal with bipolar devices in monolithic memories.   Related to the low power requirement is the high density requirement, which will be considered with the monolithic memories in particular.   However, economics in terms of process complexity or yield will not be discussed, not because of the lack of arguments for the bipolar technology, but rather because it is not such a big issue any more.   The pressure towards higher performance in MOS has induced more complex MOS processes, whereas bipolar processes have maintained or even reduced their number of steps.   To stay within this framework, the discussion will be restricted to devices and circuits that can be made with presently available processes.

2. Bipolar logic circuits

2.1. The load device problem

The very first and basic question arising from the delay–power diagram of figure 2 is: 'How can low power be realized in bipolar logic?' To understand the problem and its solution we need not consider complicated circuits.   A simple inverter suffices as a representative example.

Figure 3 depicts such an inverter both in circuit and layout representation.   It

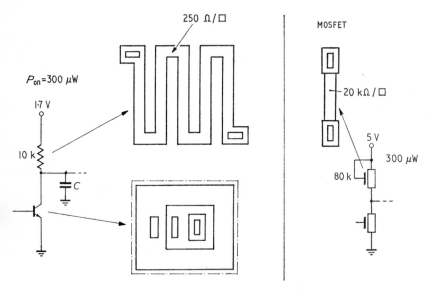

Figure 3.   Bipolar inverter at a 300 μW power level.   The load resistor consumes a relatively large area.   MOSFET loads require a smaller area.

consists of a switching transistor and a load resistor. The basic task of the latter is to provide a controlled charge current for the node capacitance C that has to be charged in order to obtain the upper logic level when the switch opens. The layout has been devised for a power level of 300 μW in the on condition of the circuit. This power corresponds just to the low power border according to our definition. With the assumed conventional 250 Ω/□ base diffusion for the resistor, its area is larger than that of the switching transistor. Of course, the situation becomes even worse when power has to be reduced further. This is a bad condition, as for large scale integration high circuit density is required and large area devices introduce parasitic capacitances that reduce speed.

The power level at which the resistor governs the circuit area depends of course on the available sheet resistance for the resistor. In comparison with the conventional bipolar circuits, the MOS device as a resistive load yields about 100 times higher effective sheet resistances, of the order of 20 kΩ/□ (Crawford 1967). Thus static MOS circuits could immediately be built for lower power than bipolar circuits before facing the area limitation, in spite of the higher supply voltage required with MOS. This is certainly one of the basic reasons why the MOSFET device has the reputation of a low power device.

With the ion implantation that is now available, sheet resistances of up to 100 kΩ/□ have been reported also for bipolar circuits (Seidel and Gibson 1971). Thus ion implantation considerably extends the power range where resistor loads are useful. However, additional process steps are required, and an extrapolation to very small dimensions again shows the area limitation even with such high sheet resistances (Hoeneisen and Mead 1972).

One possible solution to the problem is the use of switched load devices which have been used in MOS dynamic logic and in complementary MOS. If one wants to work with static loads, it is desirable to find a load device that is not subject to the difficulty of area limitation and does not require additional process steps.

2.2. *The transistor as a load device*

Fortunately, a transistor complementary to the switching transistor provides such a load device. Its function is better understood when one considers the task of the load device in a way illustrated by figure 4. Here, the conventional resistor loads are viewed as a means of distributing a total chip current I_t to the switched nodes of the individual gates to provide decoupled current sources I_0 for charging the node capacitances. A multitude of transistors with fixed emitter–base voltages fulfils the same function of current distribution as shown in figure 4(b, d). These transistors have to be complementary to the npn switching transistors. By using lateral pnp transistors no additional process steps are required in the standard buried collector process.

Seen from the collector side the transistor is a high impedance device and is thus a very good approximation to the desirable constant current source as is

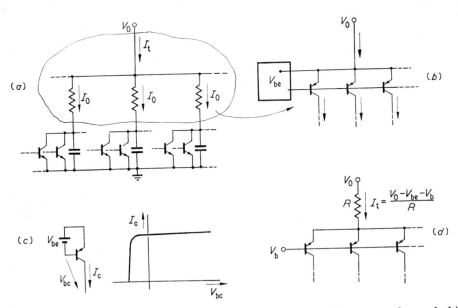

Figure 4. (*a*) Loads viewed as a means of distributing the total chip current I_t. (*b*) pnp transistor as current distributor (load device). (*c*) Constant current characteristic of the load transistor. (*d*) Current control by series resistor between supply and common emitter line.

obvious from the common base characteristic. Seen from the emitter side the transistors show a relatively low impedance and the currents can thus be varied over orders of magnitude with little V_{be} change.

The proper current level can either be adjusted via the V_{be} by a regulator or control circuit, or can be determined via a series resistor between the supply and the common emitter line (see figure 4). Instead of, say, 1000 individual high ohmic load resistors only a single low ohmic resistor is required now on the chip. Its area consumption is negligible. The individual load transistor can always be laid out with minimum dimensions at almost any practical current level. Hence, the problem of the area limitation at some low power is now solved.

This feature of the bipolar transistor as a current distributor and constant current source has been utilized in linear circuits for a long time (Widlar 1965), whereas its advantages for the load paths of digital circuits have not been well recognized until more recently. A demonstrative example of transistor loads in low power digital circuits is the frequency divider in a bipolar wristwatch circuit reported by several authors (Greuter and Korom 1971, Keller 1971, Ruegg and Thommen 1972, Keller and Kreitz 1971). Figure 5 describes its principal function.

The quartz oscillator frequency f_0 of about 16 or 32 kHz is reduced to 1 Hz

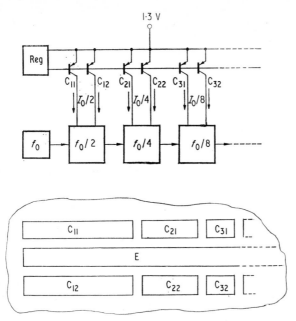

Figure 5. Current distribution by pnp transistor in wristwatch circuit.

for the stepping motor by a chain of flip-flops each dividing the frequency by a factor of 2. The two load devices for each flip-flop are realized with lateral pnp transistors as discussed with figure 4. Since all these pnp transistors have common base and emitter potentials, they can be laid out as a multicollector transistor as shown here in the layout (figure 5). According to the decreasing switching frequency in the flip-flop chain, the currents may be reduced by the same steps of 2 along the chain. This can be considered in the layout of the multicollector device by reducing the lengths of the collector pairs accordingly. Such divider chains consume less than 10 μW and it is interesting to note that with this approach the bipolar wristwatch circuits have become comparable in power economy with C-MOS wristwatch circuits of the same generation.

Another example for pnp transistor loads is a logic circuit called merged transistor logic (MTL) (Berger and Wiedmann 1972) or integrated injection logic (IIL) (Hart and Slob 1972). It will be described later in conjunction with superintegration. Figure 6, however, shows beforehand the delay–power diagram of figure 2 with this new circuit. The new line corresponds to the experimental large scale integration chips made by Hart and Slob (1972). Power dissipation is in the range of 10 μW per circuit, causing a drastic change in the picture of bipolar logic.

To summarize this section, the lateral pnp transistor as an excellent current distributor opens the way to very low power, as it liberates us from the area

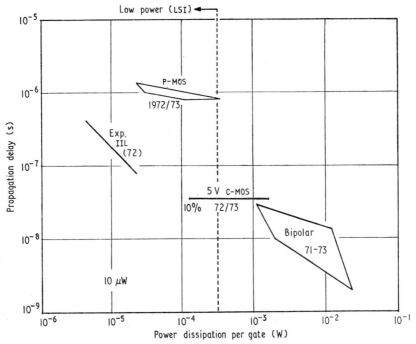

Figure 6. A pnp transistor load permits low power in bipolar logic as
IIL example (10 μW) shows.

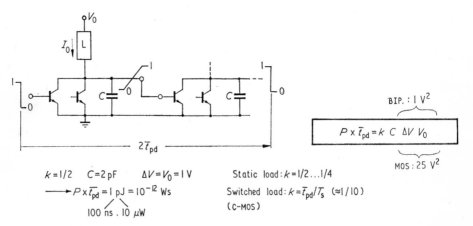

Figure 7. Figure of merit: power–delay product explained on a NOR-
gate.

limitations of the resistor load. Also the constant current feature makes it superior to resistor loads. Another very important potential of the pnp transistor load will emerge in the discussion of superintegration.

2.3. *Power–delay product*

When we compare now in figure 6 these IIL circuits with the p-MOS circuits, we see a puzzling difference in power efficiency. At the same power level the bipolar IIL circuit yields at least an order of magnitude better speed. This difference is best characterized by the power–delay product which is a well known figure of merit for the power efficiency of logic circuits (Josephs 1965) (see figure 7).

The propagation delay of a gate is caused by the ever present capacitance C. Via an idealized equivalent circuit a product of power times average delay can be derived that is given by the product of a numerical constant k, the node capacitance C being switched, the logic swing ΔV (that is, the voltage difference between the two logic levels), and the supply voltage V_0 (compare with figure 7).

For static loads the factor k lies between $\frac{1}{4}$ and $\frac{1}{2}$. As an example: with $k=\frac{1}{2}$, $V_0=\Delta V=1$V and $C=2$ pF we get 1 pJ$=10^{-12}$ Ws or, for example, 100 ns delay at a power of 10 μW. That is about what the IIL circuits showed us in figure 6.

Having the choice between circuits that can all be realized for low power, one would of course prefer the one with the smallest power–delay product as it requires the smallest power for a specified delay. Here again a comparison of bipolar with MOS logic circuits appears worthwhile.

Owing to the high transconductance of the bipolar transistor, bipolar circuits can principally be operated at much lower voltages than MOS circuits. Remember that a bipolar transistor changes collector current by a factor of 10 when V_{be} is changed only by 60 mV. In a medium delay range of 20–50 ns typical MOS circuits require about 5 V of ΔV and V_0, whereas bipolar circuits can readily be operated at about 1 V levels. The resulting difference of a factor of 25 in the voltage product of the power–delay equation cannot be overcome by the somewhat smaller capacitance of MOS devices, especially if additional load capacitances are present.

The use of dynamic logic or complementary MOS (C-MOS) solves the problem for MOS at least partly via the reduction of the k-factor. Owing to the switched load, k becomes the ratio of propagation delay t_{pd} to the switching period T_s, which can practically be of the order of 0·1. The present trend to C-MOS thus becomes quite understandable.

Complementary *bipolar* logic circuits analogous to C-MOS are not considered feasible (Josephs 1965); however, clocked dynamic logic has been reported also in bipolar technology (Grundy *et al* 1972). Anyhow, the present potential of static bipolar circuits as already illustrated by the IIL example is attractive enough to promote bipolar circuits. Hence, the following discussion will be restricted to static circuits and the potential of bipolar devices for reduction of the voltage and capacitance terms in the power–delay equation will now be shown.   But it should be pointed out here that the following considerations apply fully only to *internal* circuits of large scale integrated chips.   Going off chip, we may find rather large capacitances that we cannot influence, or a given noise level that might require larger swings and accordingly larger supply voltages.

2.4. *Schottky barrier diodes for power (–delay) reduction*

The Schottky barrier has been introduced in monolithics because it provides

(i) Negligible charge storage.
(ii) A lower forward voltage than the pn junction.
(iii) A wider range of forward voltage to choose from because of variation of
      barrier height with metal and semiconductor conduction type (Yu 1970).

At present Schottky diodes are mainly used to prevent deep saturation of transistors, so that saturation delay is eliminated (Tada *et al* 1967).   In typical low power applications with delays above 10 ns this is not so important.   There is, however, a second positive effect.   The Schottky clamped transistor has a higher saturation voltage, so that the lower logic level becomes higher and the logic swing becomes smaller.   This means a smaller power–delay product.

    Figure 8 shows another example of logic swing reduction, the Schottky diode transistor logic proposed by Schuenemann and Wiedmann (1972). As opposed to usual diode–transistor logic the diodes here are associated with the collector of the inverter transistor and not with the base.   In this way, the

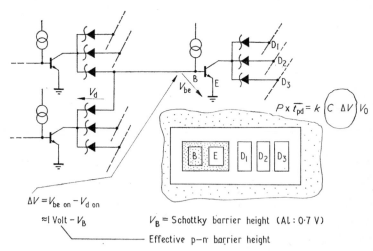

Figure 8.   Schottky diode transistor logic by Schuenemann and Wiedmann (1972)—an example of reduction of logic swing by Schottky diodes.

diodes can be integrated with the collectors of the inverter transistors as indicated in the layout. Thus capacitance and area can be kept small.

The logic swing ΔV is just the difference of the emitter–base on-voltage and the Schottky diode on-voltage. This equation can also be written in terms of barrier heights, if an 'effective barrier' for the emitter–base junction is defined. For a typical transistor this comes out at about 1 volt. With an aluminium Schottky barrier of 0·7 V the swing then becomes about 0·3 V (compare figure 8), with PtSi a swing even as low as 0·2 V is possible. Just this possibility of getting small logic swings via on-voltage differences makes the Schottky diode very attractive for low power circuits. This potential of Schottky diodes has probably not been sufficiently considered yet.

2.5. *Superintegrated devices*

The layout in figure 8 shows a structure that is a modest example of a category of devices or circuits that can be characterized as superintegrated. This term was coined by Gilbert (1970) and has been used for bipolar monolithic structures that perform the functions of several components, but do not allow a very clear separation into these individual components. The number of ohmic contacts in such a structure is typically smaller than one would expect from the individual components being represented. For instance the device structure in the Schottky diode transistor logic carries no collector contacts nor diode cathode contacts. In designing such structures 'consideration is focused on what can be achieved

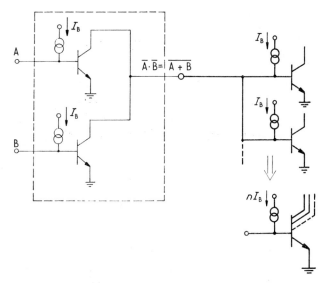

Figure 9. Example of superintegration: MTL circuit (Berger and Wiedmann 1972, Hart and Slob 1972). The derivation of a basic circuit element (multicollector inverter with base current source).

using advantageous geometries and various semiconductor mechanisms, such as minority carrier injection and collection, majority carrier modulation, local field control etc' (Gilbert 1970). Superintegration saves area and therefore capacitance. And, according to the power–delay equation, capacitance reduction means power saving.

A more complex example of superintegration, the previously mentioned merged transistor logic (Berger and Wiedman 1972) or integrated injection logic (Hart and Slob 1972) is explained in figures 9–11. In figure 9 a NOR circuit is shown that resembles the type used in conjunction with the discussion of the power–delay product (figure 7). However, the current sources are associated here with the bases of the switching transistors instead of with the collectors. This makes no difference for the circuit operation, as the input transistors of other gates are driven from the collector node and thus provide the required current source to the collector node.

For this fan-out to the next inputs a single multicollector transistor and a single base current source may be used. The transistors of the first gate can thus also be considered as being parts of such multicollector devices, or in other words any complex logic can be built out of multicollector devices having a base current source.

The first and essential trick in the realization of this device is to use inversely operated npn transistors as shown in figure 10. A common n-type substrate can be used for all emitters of the npn transistors on the chip. The usual emitter diffusion into the diffused base now forms the multiple collectors.

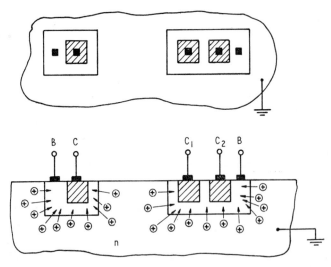

Figure 10. MTL—inversely operated npn transistor and base current by minority carrier collection.

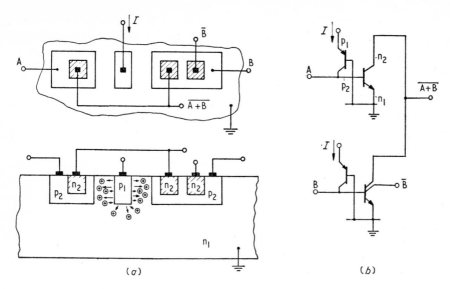

Figure 11. MTL: (*a*) carrier injection by p-emitter; (*b*) the equivalent circuit.

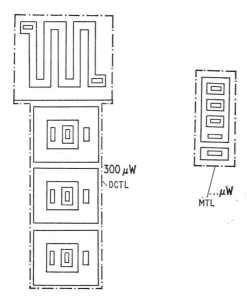

Figure 12. An area comparison of DCTL and MTL. As opposed to the DCTL, the MTL is not fixed to a certain power level.

I

The required current to the base can be attained by a collection of excess holes that are generated in the emitter region near the base.   Such excess holes can be continuously generated by injection from a p-type emitter as shown in figure 11.   This simple structure does not require many process steps.   As an equivalent circuit for this superintegrated device one can draw a pnp transistor as a load device properly connected with the multicollector npn transistor.   In the structure they are not clearly separated as it is typical for a superintegrated device.

Produced with simplified present buried collector processes, such circuits can be operated at delays above about 15 ns with a power–delay product of about 1 pJ.   That means, for example, that at a power level of only 10 μW a speed corresponding to a 100 ns delay is still possible.   This high efficiency is due to the small device capacitance resulting from the superintegration and also to the very low supply voltage.   When a constant current is supplied to all p-emitters on the chip in parallel the supply voltage on the chip is only about 0·75 volts.

Figure 12 gives an instructive comparison of an MTL with a DCTL circuit of equal logic power.   The very small MTL circuit can be supplied even at a 1 μW power level without requiring any change in size, whereas the rather large resistor of the DCTL dissipates 300 μW.   The lateral pnp transistor here not only proves to be a good load device for low power, it also lends itself to super-integration.   There is still room for improvement in the power–delay product, especially if the processes are better adapted to the new device.

2.6. *Influence of production processes*

Although this paper centres on *devices* for low power, it is appropriate here to say a few words on the processes that produce these devices.   Obviously, we have to choose or devise the right process for our low power goal.   It should provide us with the high impedance load device and other devices like Schottky barrier diodes or superintegrated devices, if we have the proper circuit to utilize them; and it should help us to minimize device capacitance.

It would take too long to discuss here all the new processes devised for high density and low power and we shall concentrate on passive isolation only (Technical Note 1971, Sanders and Morcom 1973).   It seems to have become the new standard in bipolar processes.   Passive isolation reduces junction areas and thus also capacitance.   An extrapolation of the MTL/IIL circuit has been made on the basis of such a process and by assuming that the oxide may touch the emitter–base junctions.   With conservative ground rules such as contact holes of 4 by 8 μm minimum, the power–delay product becomes about 0·2 pJ for a fan-out of 3.   This complies with the figure given by Evans *et al* (1973) for the OXIM process.

As a better illustration, the corresponding line has been introduced in the delay–power diagram (figure 13).   It shows that a 20 ns delay at only 10 μW power dissipation is within reach.

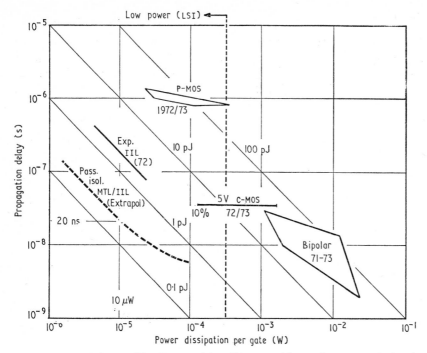

Figure 13. Power–delay diagram with MTL/IIL extrapolation for passive isolation. Good performance in spite of low power!

2.7. *Logic circuit summary*

The above considerations and examples have shown that low power bipolar logic is possible through a proper utilization of the available devices and techniques. The lateral pnp transistor as a load device opens the door to even very low power. At a specified delay power can be diminished by the reduction of circuit capacitance, logic swing or supply voltage. Superintegration is thus a low power tool, as it generally minimizes capacitance and in the case of MTL minimizes supply voltage. Schottky diodes can contribute to low power by logic swing reduction, and processes like passive isolation can reduce power through capacitance reduction. Possible power–delay figures considerably below 1 pJ destroy the notion of low power implying low performance.

3. **Bipolar monolithic memories**

Now, the problem of low power and high density in bipolar monolithic memories shall be considered. We shall restrict the discussion to the most important high density random access read/write memories.

3.1. Orientation

A random access memory chip usually consists of a regular array of memory cells surrounded by some support circuitry. The latter consists mainly of logic circuits like address decoders. The advantage of bipolars in the logic field in terms of power utilization has just been underlined. Even MOSFET memories often use bipolar support circuits because of their advantages.

The main difficulty in bipolar memories has been to realize memory cells that are very small and which at the same time dissipate little power. Therefore, bipolar memory chips have been lagging behind MOSFET memories in a density distance of a factor of about 4. At present 4000-bit MOSFET memory chips are being introduced as a new industry standard (Wolff 1973) and an account of an experimental 8000-bit chip has been published (Hoffman and Kalter 1973) whereas the most dense bipolar memory chips in production offer 'only' 1000 bits/chip (Baker *et al* 1973). Since bit density is strongly coupled with the price per bit, it is understandable that bipolar memories have been applied mainly where the speed of MOSFET memories has been insufficient, as in high speed buffer memories or in main memories of fast computers.

To answer the question whether bipolar devices can fulfil the requirements of a high density memory cell after all, we first have to have some orientation figures. For estimating the power dissipation limitations we assume that we want to pack 10 kbit/cm² of package surface (figure 14). With the figures of 100 mW and up to 300 mW of possible heat removal mentioned before, we then arrive at a power dissipation per bit of the order of 10 μW to 30 μW. Although parts of the peripheral support circuits on the chip or package can be switched

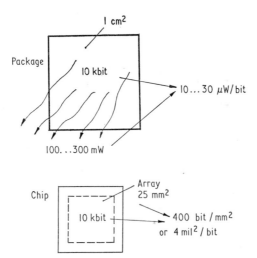

Figure 14. Power and density orientation figures for high density memory cells.

off in the standby state, we have to leave some power at least for the then necessary switches. Thus it appears reasonable to demand storage cells having less than 10 µW of power dissipation or less than 4 µA of current assuming a 2·5 V supply.

For the density requirement one gets a feeling in numbers that are easy to remember, if one assumes an available chip area of 5 mm by 5 mm for the 10 000 bits. The corresponding bit density is $400/mm^2$, or in other words the area per cell is 4 mil², if one uses the obsolete mil units that are convenient in this case.

3.2. Discussion of the flip-flop memory cell

The basic tools for achieving low power and high density memory cells are the same as for low power logic. We will discuss this using the flip-flop shown in figure 15. The symmetric flip-flop cell consists of two cross-coupled switching transistors (only one of them can conduct at a time), two load devices and two switches for coupling to the read/write line pair.

Resistor loads cannot be used at this low power and the high density required. Also the higher resistivity of the MOSFET load is insufficient. Therefore, MOSFET memories above 1000 or 2000 bits/chip utilize dynamic storage cells exclusively (Altman 1972, Wolff 1973). There are also proposals for bipolar dynamic storage cells (Mar 1971, Henn 1971). However, in a similar way to the logic circuit field, the potential of bipolar devices for static circuits has not yet been fully utilized and therefore we shall confine ourselves here to static stable storage cells only.

The pnp transistor is again an excellent choice for the load device. Flip-flops

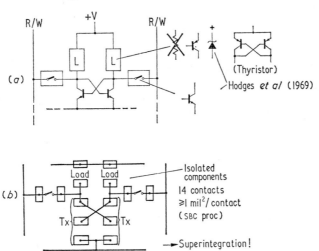

Figure 15. (a) Principal flip-flop memory cell: circuit and devices used for low power. (b) Area estimate with isolated components clearly shows the necessity of superintegration to reach the 4 mil² goal.

with this load have been found stable down to a few nanowatts of power (Wiedmann and Berger 1971).   Also reverse-biased Schottky diodes have been proposed by Hodges *et al* (1969); however, the operation current reported so far is too high (25 μA with a threefold increase with warm-up).   With the pnp transistor one is more flexible.   Also cross-coupled transistors complementary to the other transistors work well; by superintegrating these transistors—npn with pnp—one gets the cross-coupled thyristor of Jutzi and Schuenemann (1972).

For the read/write line switches, either the emitter–base junction of a transistor or a diode is used.   Here, a Schottky diode is an excellent choice in all cells, where it can be superintegrated into the n⁻ region of another cell component.

When we count the number of contacts that we get by using isolated active components (there are 14 contacts) we soon learn that superintegration is a necessity.   Isolated devices in standard buried collector technology have an area per contact of about 1 mil².   That makes a silicon area of 14 mil² for the 14 contacts of the cell and possibly more because of interconnection lines.

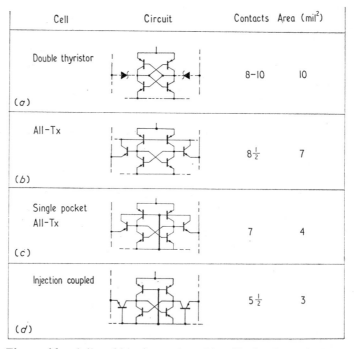

Figure 16.   A list of bipolar static stable cells that have the low power, high density potential, demanded in figure 14.   The areas are those for the standard buried collector process and 5 μm line width; passive isolation reduces the cell size by about a factor of 2.   (*a*) Double thyristor (Jutzi and Schuenemann 1972, Dunn *et al* 1973); (*b*) all-transistor (Wiedmann and Berger 1971); (*c*) single pocket all-transistor (Wiedmann and Berger 1972); (*d*) injection coupled (Wiedmann 1973).

Even with passive isolation we would have about 5 mil², still more than our 4 mil² goal and, particularly for the passive isolation case, one has to consider additional area requirements for interconnections.

Superintegration reduces the number of contacts and thus possibly the area requirement. Figure 16 shows a table of superintegrated cells that can be operated at less than 10 μW of power. The thyristor as the oldest super-integrated device needs 8–10 contacts in the cross-coupled arrangement. The reported cell area lies around 10 mil² in the standard buried collector process. With passive isolation all cell sizes reduce by a factor of about two, so that this cell is a contender for the high density memories. The other cells—all with pnp load devices—offer even higher densities according to the decreasing number of contacts. The injection coupled cell (Wiedmann 1973) has only $5\frac{1}{2}$ contacts. This cell will be described shortly, as it is from a device point of view particularly interesting.

3.3. Injection coupled memory cell

The operation of this cell (figure 17) is based on the injection principle of MTL and IIL. When current I_0 is supplied to the p_1 region, holes are injected into the n^- region. They are collected both by p_2 and p_3. If the junction p_2n^- is turned on due to the collected positive charge, an electron injection from the buried layer into p_2 starts. Thus the inverse npn transistor turns on and withdraws the current from p_3. Hence, the other transistor is kept off. The reverse state is possible as well, so that we have a flip-flop. The p region of the

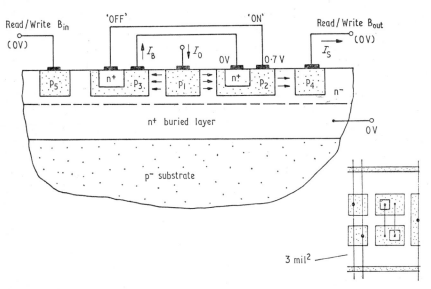

Figure 17. Injection coupled memory cell showing cross-section and principal layout (Wiedmann 1973).

on-transistor, here p_2, reinjects holes into the n^- region and they can be collected by p_4. Thus the state of the cell can be sensed by the current flowing out of p_4 or p_5. The flip-flop may be considered here as a switch that can steer the injection current of p_1 either to p_4 or p_5. Writing is performed by injecting holes from p_4 or p_5 to turn the adjacent transistor on. As shown earlier, cell size is 3 mil^2 in the standard buried collector process. Passive isolation allows a reduction of down to about 1 mil^2. Hence, this static stable cell offers the same density as MOSFET one-device cells do, but at a ten times better speed.

4. Conclusion

The present spectrum of the supply of digital circuits strongly suggests that low power dissipation cannot be achieved with bipolar devices. However, such a conclusion has been shown to be a fallacy. Bipolar digital circuits can move very deeply into the low power region by utilizing common base load transistors instead of the conventional resistor load. Moreover, at a given power level, bipolar circuits have a higher speed potential than MOSFET circuits. This potential can be utilized by logic swing reduction through Schottky diodes, and by supply voltage and capacitance reduction with superintegration. Superintegration is especially important in the monolithic memory field with its high density requirements. Static stable superintegrated memory cells are now known that offer densities comparable with those of dynamic MOSFET one-device cells (eg Stein and Friedrich 1973, Proebsting and Green 1973) and provide sufficiently low power.

As stated before, the bipolar low power potential has not been exploited commercially. There is, however, one exception: bipolar implementations of wristwatch circuits are quite successful in Europe. It would be beyond the scope of this paper to try a comprehensive analysis of the reasons why bipolars are actually lagging in the low power field, but let us consider at least two possible reasons that might have contributed.

First, confidence in bipolars has been very low because of an underestimate of their potential. It is hoped that this paper conveys some confidence by showing their actual potential. Secondly, as the example of superintegration might illustrate, realization of this potential requires a very close cooperation of technologists, device researchers and circuit designers. This is a challenge particularly to the device researcher who stands in the centre of this chain.

References

Altman L 1972 *Electronics* **45** 63–77
Baker W D *et al* 1973 *Electronics* **46** 65–70
Berger H H and Wiedmann S K 1972 *IEEE J. Solid St. Circuits* **SC-7** 340–6
Crawford R H 1967 *MOSFET in Circuit Design* (New York: McGraw-Hill) p 87
Dunn R *et al* 1973 *ISSCC 73, Digest of Technical Papers* (New York: Lewis Winner) pp 66–7

Evans W J *et al* 1973 *IEEE J. Solid St. Circuits* **SC-8** 373–80

Gilbert B 1970 *ISSCC 70, Digest of Technical Papers* (New York: Lewis Winner) p 120

Greuter A and Korom A 1971 *Neue Züricher Zeitung (Technische Beilage)* No 15 S13–14

Grundy D L *et al* 1972 *Microelectronics* **4** No 3, 10–30

Hart K and Slob A 1972 *IEEE J. Solid St. Circuits* **SC-7** 346–51

Henn H H 1971 *IEEE J. Solid St. Circuits* **SC-6** 297–300

Hodges D A *et al* 1969 *IEEE J. Solid St. Circuits* **SC-4** 280–4

Hoeneisen B and Mead C A 1972 *Solid St. Electron.* **15** 891–7

Hoffmann W K and Kalter H L 1973 *ISSCC 73, Digest of Technical Papers* (New York: Lewis Winner) pp 64–5

Josephs H C 1965 *Microelectronics and Reliability* **4** 345–50

Jutzi W and Schuenemann C H 1972 *IBM J. Res. and Develop.* **16** No 1, 35–44

Keller H 1971 *Radio Mentor Elektronic* **37** No 7 S420–6

Keller H and Kreitz W 1971 *Elektronik* **20** No 8 S261–4

Mar J 1971 *IEEE J. Solid St. Circuits* **SC-6** 280–3

Proebsting R and Green R 1973 *ISSCC 73, Digest of Technical Papers* (New York: Lewis Winner) pp 28–9

Ruegg H W and Thommen W 1972 *IEEE J. Solid St. Circuits* **SC-7** 105–11

Sanders T J and Morcom R 1973 *Electronics* **46** 117

Schuenemann C L and Wiedmann S K 1972 *IBM Technical Disclosure Bulletin* **15** 509

Seidel T E and Gibson W C 1971 *Int. Electron Dev. Meeting, Washington 1971* (New York: IEEE) paper 17.3

Stein K U and Friedrich H 1973 *ISSCC 73, Digest of Technical Papers* (New York: Lewis Winner) pp 30–1

Tada K *et al* 1967 *Proc. IEEE* **55** 2064–5

Technical Note 1971 *Microelectronics and Reliability* (London: Pergamon) **10** 471–2

Widlar R J 1965 *IEEE Trans. Circuit Theory* **CT-12** 586–90

Wiedmann S K 1973 *ISSCC 73, Digest Technical Papers* (New York: Lewis Winner) pp 56–7

Wiedmann S K and Berger H H 1971 *IEEE J. Solid St. Circuits,* **SC-6** 283–8

—— 1972 *Electronics* **45** No 4 83–6

Wolff H 1973 *Electronics* **46** No 8 75–7

Yu A Y C 1970 *IEEE Spectrum* **7** No 3 83–9

Display devices

S van Houten

Philips Research Laboratories, Eindhoven, Netherlands

Abstract

An electronic display converts electrical signals into visual information. The physical phenomena used to realize the display function must satisfy requirements which arise both from the electronic circuitry needed to control the display and from considerations of visual perception. Some of the basic requirements will be discussed, and the discussion illustrated by the use of liquid crystals and of electrochromic compounds. Liquid crystals form a very interesting class of materials with peculiar properties. They show several electro-optic effects that may be used to advantage for display applications. The novel electrochromic display makes use of a colour change of an organic substance due to a charge transfer at the electrodes. Finally, we shall compare liquid crystal and electrochromic displays with two other well known displays, utilizing light-emitting diodes and gas discharges respectively.

1. What is a display?

An electronic display converts electrical signals into visual information, comprehensible to the human observer. The information is usually presented as alpha-numeric symbols and/or graphics. An extensive introduction to displays has been given by Luxemberg and Kuhn (1968).

Displays may be of all sizes. Nowadays the two classes of displays most frequently used are large displays with high information density, as in television and in computer output applications, and small numerical indicators for instruments, calculators and clocks. This is illustrated in table 1, where displays are

Table 1. Displays

Approximate size (in characters/display)	1–16	56	256	1000	Bigger
Main applications	Clocks Watches Calculators Instrumentation Indicators Keyboard displays Automotive	Mini-computers Instrumentation Keyboard displays Bar charts Avoid radar Cockpit display Large characters	Computers Control systems Banking Reservation systems Patient monitoring	Computers Control systems Graphics Editing Videophone	Graphics Editing TV

divided into five categories: alpha-numeric indicators for up to, say, 16 characters; displays for about 56, 256 and 1000 characters; and bigger ones. It is to be expected that in the near future intermediate-size displays for the presentation of, say, 20 to 600 characters will be used more often than now, especially in educational, control and information systems.

There are several possible formats for representing digital information. This is shown in figure 1, where three basic formats are indicated, namely preformed characters, segmented characters and dot-matrix characters.

Preformed characters are used in the well known indicator tube. This is a gas-discharge tube in which a number of preformed cathodes are stacked one behind the other. This simple method is suitable for a restricted number of symbols only.

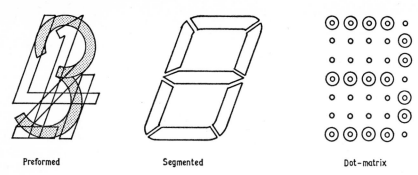

Preformed   Segmented   Dot-matrix

Figure 1.   Possible formats of symbols.

Segmented characters are widely used, too.   Especially when less than, say, 30 characters have to be displayed.   The minimum number of segments is 7 for the display of numerals, but larger for an alpha-numeric display.

A dot-matrix has the advantage that—thanks to the larger number of display elements—readability is much better and the chance of misreadings is smaller than in the two other methods.   This may be important in those applications where, for legal reasons, a foolproof presentation is required.   Moreover, a dot-matrix is the only practical solution for big displays.   More driving circuitry, however, is needed.

2. Aspects of displays

Let us now see which factors determine whether or not a display is good, restricting ourselves to two aspects only:

(i) Driving and decoding circuitry: how simple (or how complicated) is the electronic circuitry required to control the electro-optical effect that is used?
(ii) Visibility: how well can the information be read?

2.1. Driving and decoding circuitry

To understand the importance of the driving and decoding circuitry it is helpful to realize what a display really does.   An electronic display converts the incoming electrical signal into visual information (figure 2).   It consists of two basic parts: the visual display device, and the peripheral electronic circuitry. These two parts are closely interdependent and the one should therefore not be developed without the other.

The electrical information at the input is usually in a binary coded form, at normal logic power levels.   These signals have to be transformed into a form suitable for the display device.   This is done in the decoding and driving

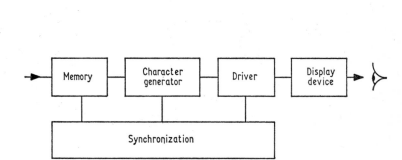

Figure 2. A display converts the incoming electrical signal into visual information.

circuitry. Here we may distinguish the following parts: memory, character generator, drivers, synchronization circuits.

The memory temporarily stores the information that is presented to the display. This is necessary because neither the speed nor the form in which the information is presented is as a rule suited to the capabilities of the display device. The character generator decodes the incoming signal into a signal that is suitable for the format of the symbols to be displayed. The input signal may, for example, be in 6-bit binary form for a choice from 64 characters. The signals coming from the character generator are still at normal logic power levels, which usually differ from the currents or voltages needed by the electrical-to-optical transducer. The drivers supply currents or voltages at the required levels. The synchronization circuit ensures that everything happens at the right moment.

In the display device itself two functions may be distinguished:

(i) selection of the individual elements;
(ii) production of light (in a passive display, modulation of light).

In a cathode ray tube, for example, both functions are performed by the scanning electron beam, the second function in combination with the cathodoluminescent phosphor screen. In other types of displays the two functions may be either combined or separate.

In a simple display consisting of a small number of individual display elements, each element may be separately connected to the driving and decoding circuitry. The selection function is then performed by switches, one for each element. For a large display, consisting of thousands of elements, this solution is obviously impracticable.

In a large array of many elements one therefore uses crossbar address (figure 3). The elements are located at the intersections of two sets of mutually orthogonal electrodes. For the selection of $n \times m$ elements only $n + m$ electrodes are needed.

Now let us see what the consequences are of using crossbar address. Selection of an element is obtained by coincident address of two mutually perpendicular electrodes. We note that all the electrodes are interconnected in a complicated way via all the display elements. Therefore the physical phenomena to be used require an input threshold below which the display element is not activated, otherwise the elements not selected will be activated at half the intensity level of the selected elements. In the absence of an inherent threshold the electronic circuit must provide that function at each individual element, but this will inevitably complicate the circuit.

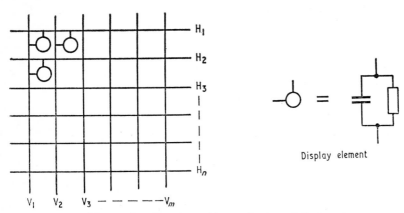

Figure 3. Crossbar address of array of display elements.

Several schemes have been proposed for the amplitude and the polarity of the voltages to be applied to the two sets of electrodes in a crossbar array. Some useful alternatives of the commonly employed 1 : 1 scheme are the 1 : 2 and the 1 : 3 voltage schemes (see Gooch and Low 1972).

The individual elements in a crossbar array can either be addressed one at a time (point sequential) or a line at a time (line sequential). In either case the individual elements are activated only for a fraction of the time, determined by the so-called duty factor F. Duty factor here means the time that an element is 'on', divided by the time between successive addresses. In order to yield sufficient light output an element should be able to give a peak output that is $1/F$ times as high as the stationary output. This is not always possible and may restrict the applicability of certain physical effects.

Sometimes each element has a memory function of its own, so that, having been addressed, it stays 'on' until a second address pulse switches it 'off'. This has two advantages: there are no duty factor problems, and the memory function to be provided by the electronic part of the display can be much simpler.

A third factor that determines whether or not a certain effect can be used in dynamically driven displays is the speed of response. It will be clear that the response time should be smaller or at least comparable to the address time. This requirement is less severe if the electro-optical effect has integration properties. In that case a number of successive address pulses has a similar effect as one pulse of the same total length.

Similar considerations as for crossbar address of a large array apply to multiplexing of a number of digits in a register display.

2.2. Visibility

The second important aspect of displays is the visibility. Many factors influence the visibility of displays, such as contrast, brightness, colour, resolution and definition. The relationships between these factors are complex and the visibility is further greatly influenced by the ambient light level.

It is useful to distinguish two ways in which information can be made visible. A display can be either passive or active. In an active display device, light is generated and the intensity is varied locally. In a passive display no light is generated, but the ambient light is modulated by, for example, absorption, birefringence or scattering.

Both methods have their own merits and disadvantages:

Active   (i) easily readable in dimly lit surroundings;

(ii) high contrast ratios possible, but at moderate ambient light levels only;

(iii) perceptional problems may arise (adaptation; flicker);

(iv) light has to be generated, which means that much electrical power is needed. Therefore, luminous efficacy is important.

The best known active display is the cathode ray tube. It is likely that, for a long time to come, this will remain the best solution in many applications that demand large displays. But there is increasing interest in other solutions such as, for example, gas discharge and injection luminescence displays.

Passive   (i) fairly readable in bright surroundings;

(ii) contrast independent of ambient light level, but not readable in the dark;

(iii) no adaptation problems; quiet presentation;

(iv) no light is generated, so that very little power is required.

To date there are still very few passive displays commercially available, but many principles have been proposed. Some of the most promising are those in which liquid crystals or electrochromic substances are used.

Now we come to a discussion of liquid crystals displays as a first example of an interesting class of electro-optical phenomena that may be used for display applications. We shall see if and how they meet the requirements described above. Later on in the paper some other types of displays will be discussed.

3. Liquid crystals

Some organic substances, apart from the three normal aggregation states, possess a fourth intermediate state between the solid and the liquid state (see figure 4). This intermediate state, usually called the mesomorphic state of the liquid crystalline state, is characterized by a high degree of long-range orientational ordering of the molecules. We may distinguish three basic types of

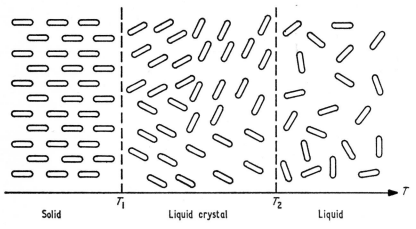

Figure 4. Liquid crystalline state as a fourth phase.

ordering, namely smectic, nematic and cholesteric, but we shall here consider only the two last named types. The basic properties of liquid crystals have been reviewed by Fergason (1964) and by Chistiakov (1967). The various electro-optic effects in liquid crystals have been summarized by Sussman (1972).

3.1. *Nematic liquid crystals*

In the nematic liquid crystalline state the anisotropic molecules are more or less oriented in parallel. This molecular ordering can be described by means of a vector, the director **n**, representing the average orientation of the molecules (see figure 5). Because of thermal motion the long axes of the molecules deviate statistically from the preferred direction. The centres of gravity are distributed isotropically.

Nematic molecules often have a structure, as shown in figure 6. Two aromatic rings are joined by a bridging group. The p-substituents of the rings

K

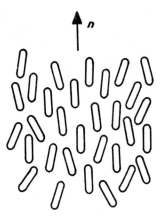

Figure 5.   Molecular ordering in the nematic state. n is the director.

may be the same or different.   The molecules are long and fairly rigid, and they have a larger polarizability along the molecular axis than perpendicular to it. The molecules often, but not always, have a dipole moment.   Some relations between molecular structure and liquid crystalline behaviour have been discussed by de Jeu and van der Veen (1972).

Owing to the orientational ordering a nematic liquid crystal is birefringent, with a very large value of Δn, a typical value of Δn being 0·2 for a nematic compound, compared with a value of Δn of 0·01 for quartz.   The anisotropy in the dielectric constant may be of either sign, depending on the presence of a dipole moment and on the angle which it makes with the molecular axis.   We shall later see that the sign of $\Delta \epsilon = \epsilon_{\parallel} - \epsilon_{\perp}$ is a very important parameter which determines whether or not certain materials show a particular opto-electronic effect.   The electrical conductivity is also anisotropic, but here generally $\sigma_{\parallel} > \sigma_{\perp}$.

In a large sample the director varies from one region to another.   Such a sample therefore is not transparent but it scatters the incident light.

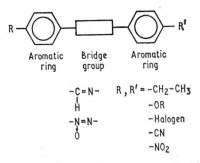

Figure 6.   Possible structure of nematic molecules.

The anisotropic properties are especially marked in 'monocrystalline' samples. Such a sample may be made in the form of a thin nematic layer of, say, 20 μm in thickness between two glass plates. We can distinguish two cases (see figure 7), namely the planar texture and the homeotropic texture.

The first-named texture can be obtained by scratching the glass plates at the inner sides or by a chemical treatment or by a combination of both methods (see Kahn *et al* 1973). The molecules align themselves along the direction of rubbing. The second texture can be obtained in a very clean cell, especially when a little surface-active substance is added to the nematic liquid crystal. Both 'monocrystalline' layers are clear and transparent.

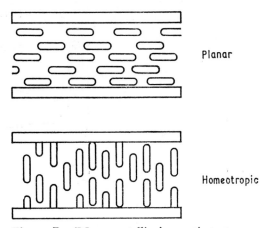

Planar

Homeotropic

Figure 7. 'Monocrystalline' nematic textures.

The perfect orientation of the molecules in either texture may be changed or disturbed by the application of a voltage across the layer. Depending on the properties of the nematic liquid crystal and on the magnitude of the voltage, several interesting effects can be observed. Two types of liquid crystal effects may be used for display applications. The first type is induced by a weak electric current, an example being the 'dynamic scattering'. The other type is caused by an electrical field, examples being the 'twisted nematic', the 'electrically controlled birefringence', and the 'cholesteric-to-nematic phase transition'. A last effect, the 'cholesteric memory' effect, is partly a current and partly a field effect.

3.2. Dynamic scattering

The best known effect is the 'dynamic scattering' (Heilmeyer *et al* 1968). Owing to the anisotropic electrical conductivity and the negative anisotropy of the dielectric constant, a current through the cell gives rise to instabilities ultimately leading to a turbulent flow of the nematic liquid crystal in the applied field (see Orsay Liquid Crystal Group 1971). The layer is then no longer transparent, but scatters the incident light strongly.

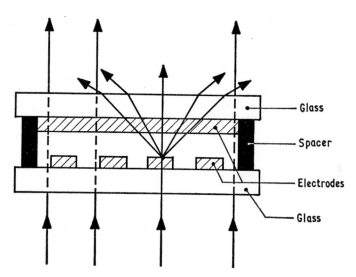

Figure 8.   Dynamic scattering cell.

The construction of a dynamic scattering cell is shown in figure 8.   Two glass plates, separated by a 20 μm spacer, are sealed together.   The seal should be hermetic, because liquid crystal materials often react with water.   Transparent electrodes are applied by evaporation or similar methods on the inner sides of the glass plates.   If these electrodes are arranged in a regular pattern, for example in a 7-segmented format, the display of numerals is possible.   The liquid crystal material must be doped such that the resistivity is 10^9–10^{10} Ω cm.

Let us now see how far the dynamic scattering effect meets the display requirements (compare table 2).   The effect can be used to modulate incident light, either that of a special light source or the ambient light.   It will therefore give a passive display.   The effect is induced by a very weak electric current, of the order of a mere 10 μA cm^{-2}.   The magnitude of the voltage to be applied is of the order of 20 V.   To avoid any electrolysis a low frequency AC voltage is normally used.

The contrast is not very high and depends on viewing angle, and because there is no clear threshold the discrimination ratio of the effect is poor.   Also, small duty cycles cannot be achieved because the effect is rather slow (cf table 2).

Figure 9. 9-digit dynamic scattering display with drive ICs.

facing page 140)

Table 2. Liquid crystal effects

Effect	Dynamic scattering	Electrically controlled birefringence	Twisted nematic	Chol. to nem. phase change	Cholesteric memory
Electrical	current	field	field	field	current field
$\Delta\epsilon$ (molecules)	neg.	neg. or pos.	pos.	pos.	neg.
Applied voltage	20 V	5 V	$\leqslant 5$ V	100 V	20 V
Current	10 μA cm^{-2}	1 μA cm^{-2}	1 μA cm^{-2}	1 μA cm^{-2}	1 μA cm^{-2}
Response time 'on'	10–20 ms	10 ms	1 ms	30 μs	30 ms
Response time 'off'	100–200 ms	200 ms	200 ms	100 μs	1 s
Threshold	poor	yes	yes	yes	yes

Moreover, the dynamic scattering quickly becomes saturated. Large display panels using crossbar address cannot therefore be made unless a memory function can be introduced in the display elements. This is, in principle, possible as will be discussed below (cf memory effect).

Multiplexing of a number of digits in a register display is also difficult although tricks have been proposed (such as a two-frequency address) to overcome these problems (see Wild and Nehring 1971, Stein and Kashnow 1971). One of the advantages of multiplexing would be the smaller number of interconnections, compared to that of a statically driven display. Another approach to reducing the number of external interconnections has been used by Gerritsma and Lorteye (1973). A buffer memory in the form of an IIL shift register is mounted directly on the glass plate carrying the electrode pattern. This is shown in figure 9. The number of external interconnections for a 9-digit statically driven display is thereby reduced from 64 to only 5.

The main advantages of a dynamic scattering display are the simple technology and the low power consumption.

3.3. Electrically controlled birefringence (ECB)

A first example of a field effect is the 'electrically controlled birefringence' (Schiekel and Fahrenschon 1971, Assouline *et al* 1971, Kahn 1972, Soref and Rafuse 1972). The effect can be understood on considering a perfectly homeotropic layer (figure 10). Such an oriented layer behaves optically as a uniaxial crystal, with the optical axis parallel to the incident light. If placed between crossed polarizers the light is completely blocked. In figure 10 the effect of an applied electrical field is also indicated. The molecules, which in this case have a negative anisotropy of the dielectric constant, are tilted through a certain angle, depending on the magnitude of the applied voltage. This means that a bire-

K§

fringence is induced locally. In transmission, therefore, interference colours can be observed. The colours, however, are not saturated.

This birefringence has a much more pronounced voltage threshold than the dynamic scattering (cf table 2). However, no use can be made of this property, as the response times are very long near the threshold, typically 2 seconds. A device utilizing this effect is, therefore, usually switched between two voltages above the threshold, corresponding with two colours (Assouline *et al* 1971). Because the ECB is slow, use can be made of the integration effect.

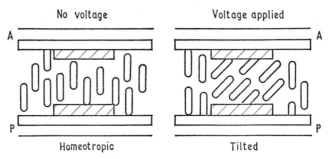

Figure 10. Principle of electrically controlled birefringence cell.

The viewing angle is strongly restricted because the effect, of course, depends on the angle of incidence of the incident light. It is also sensitive to temperature and thickness variations. Applications, therefore, may be found in the field of projection display. Furthermore, it is a field effect, with low values of the applied voltage. A typical value is 5 V. In order to prevent any dynamic scattering one usually applies an AC voltage with a frequency above the critical frequency for dynamic scattering. This frequency lies in the audio range.

It is also possible to observe an electrically controlled birefringence effect starting from a planar texture (see de Jeu *et al* 1972). In that case the nematic molecules must have a positive anisotropy of the dielectric constant.

3.4. Twisted nematic

A second example of a field effect is the 'twisted nematic' (Schadt and Helfrich 1971, Leslie 1970). A device involving this effect can be made in the following way. We start with two glass plates treated to induce the planar texture. Before the two glass plates are sealed together, one of them is rotated through 90° with respect to the other. A display cell is shown in figure 11. Near the upper glass plate the molecules are oriented parallel to the plane of the drawing; near the lower glass plate they are perpendicular to it.

As the wavelength of light is less than the thickness of the sample, the gradual rotation of the molecules across the layer leads to a rotation through 90° of the

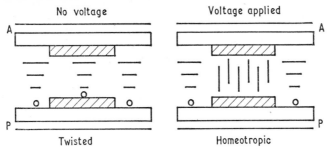

Figure 11. Principle of twisted nematic display cell.

plane of polarization of incident plane-polarized light. When such a cell is placed between two parallel polarizers no light will be transmitted. If the anisotropy of the dielectric constant $\Delta\epsilon$ of the nematic molecules is positive, an electric field will rotate the molecules until they are parallel to the field. Then there is no longer a rotation of the plane of polarization and the layer is transparent between parallel polarizers. As soon as the applied field is switched off the molecules reorient themselves to their original twisted orientation. This happens because the molecules in contact with the two glass plates never did rotate and now force their neighbours to take up their original orientation.

This effect has several interesting features (cf table 2). The contrast that can be obtained is higher than that of the other two effects, although it also depends on the viewing angle. It is a field effect, with an operating voltage of the order of 5 V or less, depending on the material used. The effect has a threshold (Gerritsma *et al* 1971, van Doorn 1973), so that the discrimination ratio is good, even better than that of the electrically controlled birefringence. On the other hand the effect is also slow. This is a promising effect for display applications.

3.5. *Cholesteric liquid crystals*

A second type of the liquid crystalline state is the cholesteric mesophase. The cholesteric mesophase should be considered as a special case of the nematic ordering. This is clearly seen when a small amount of an optically active substance is added to a nematic liquid crystal. Then a twist is induced in the nematic structure similar to, but much stronger than in the 'twisted nematic'. The same occurs if the molecules themselves have an asymmetric carbon atom as is the case in cholesterol derivatives. The cholesteric structure can schematically be described to consist of layers, as is shown in figure 12. In each layer the ordering is nematic, but the molecules are rotated from layer to layer through a small angle. After a large number of layers, say 1000, the orientation is rotated through 180°. There is a screw axis with a pitch P, along which the orientation of the molecules changes gradually. P can be of the order of optical wavelengths, depending on the molecules.

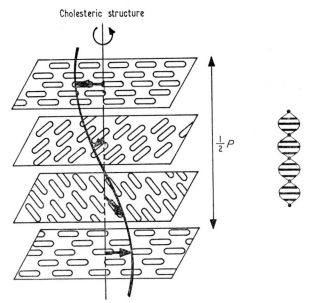

Figure 12. Cholesteric structure.　$0 \cdot 1 \ \mu m < P < 2 \cdot 0 \ \mu m$.

In a thin cholesteric layer two different cholesteric textures, the focal conic and the planar texture, may be found (figure 13). In the focal conic texture the molecules lie in planes that are perpendicular to the glass surface, but the orientations of the screw axes are randomly distributed. A cholesteric layer having this texture is not transparent but scatters light. In the planar texture the screw axes are oriented perpendicular to the glass surface and hence the molecules lie in planes parallel to the glass surface. A planar cholesteric layer is transparent.

3.6. Cholesteric-to-nematic phase transition

One of the effects of the cholesteric mesophase that may be used for display purposes involves the 'cholesteric-to-nematic phase transition' (Wysocki *et al* 1969, Jakeman and Raynes 1972).

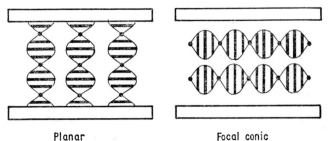

Planar　　　　　　　　　　　　Focal conic

Figure 13. Textures of cholesteric layers.

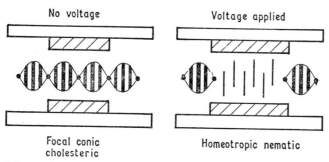

No voltage   Voltage applied

Focal conic   Homeotropic nematic
cholesteric

Figure 14. Principle of cholesteric-to-nematic phase change.

If we place a focal conic cholesteric sample in an electrical field it is possible to unwind the helical structure so that a nematic structure is formed (figure 14). In the focal conic state the incident light is scattered because of the imperfect ordering of the helices. In an electrical field the molecules all become oriented parallel to the field, so that a clear homeotropic nematic texture is formed. For this effect to occur, the molecules should have a positive anisotropy of the dielectric constant.

It is a field effect with a threshold. The speed of response can be made high by applying a high voltage so that the effect may be suitable for crossbar address. Typical values for the applied voltage are up to 100 V, which makes direct drive from ICs questionable (cf table 2). The short decay time implies that no use can be made of the integration effect.

3.7. *Memory effect*

A second effect involving cholesteric liquid crystals makes use of a mixture of a nematic and a cholesteric substance. Such a mixture has cholesteric properties, but the pitch P is large. The anisotropy of the dielectric constant has to be negative. The memory effect is based on the transition from the planar to the focal conic texture (figure 15). Application of a DC or a low-frequency AC

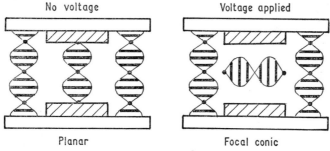

No voltage   Voltage applied

Planar   Focal conic

Figure 15. Principle of memory effect.

voltage of about 20 V causes a small current to flow in the sample, which induces a dynamic scattering effect.   When the voltage is switched off the sample remains in the strongly scattering, focal conic state.   This focal conic texture is meta-stable; hence a display involving this effect possesses a memory function.   Once written in, the information can be stored for months (Heilmeyer and Gold-macher 1968, 1969).

The information can be erased by the application of a high frequency AC voltage which in turn induces a transition to the clear, planar texture.   This transition is a field effect and is rather slow.   The erase voltage strongly depends on the value of the anisotropy $\Delta\epsilon$.   For MBBA/EBBA it is about 100 V, for other substances it can be lower (de Jeu and van der Veen 1973).   The effect has a threshold; the discrimination ratio is reasonable.   The writing time is compar-able to that of the dynamic scattering, the erase time is of the order of 1 second. The main advantage is the memory function, which permits crossbar addressing and low duty cycles (cf table 2).

3.8. Comparison of the liquid crystal effects

In table 2 the most pertinent facts about the liquid crystal effects have been collected.

All liquid crystal displays have a very low power consumption and mostly low switching voltages.   The technology used to make the devices is flexible and potentially cheap.   Devices of many sizes can be made, using the same tech-nology.   Some field effects have sharper thresholds, so that the discrimination ratio is better than that of the 'dynamic scattering' effect.   This is important in connection with the possibility of multiplexing a number of digits.   A second important parameter in this respect is the speed of response, which for most effects is too low for crossbar address.   On the other hand long response times facilitate the use of the integration effect.

An important aspect of liquid crystal displays is life.   Life problems may result either from decomposition of the substances or from decreasing perfection of the zero texture or from both.   It has been proven that a life of more than 10 000 hours is possible for AC driven 'dynamic scattering' displays.   Much less is known about the field effects, though it is to be expected that life is at least as good as far as chemical decomposition is concerned.   The perfectness of the texture, however, is much more critical, but little is known about this problem.

Liquid crystals can be used in a restricted temperature region only.   Now-adays several mixtures can be obtained with a fairly large temperature region, for example from -20 to $+80\,°C$.

Liquid crystal displays can be operated either in the transmissive or in the reflective mode.   In the former case a lamp is usually needed for illumination of the display.   In the latter case no artificial illumination is required and the dis-play is passive.   Such a display can be used in bright sunlight without loss of contrast.

4. Electrochromic display (ECD)

An entirely different type of display is the electrochromic display. Several electrochromic displays have been suggested in the past and have been described in the literature. They are mostly based either on a selective precipitation of a metal (eg Zaromb 1962), or on the formation of colour centres in metal oxides (eg Alburger 1957) or on a local variation of the acidity of a solution leading to a colour change of an acidity indicator (eg Deb 1969). Such displays, however, have several disadvantages such as low speed, poor reversibility and high power consumption.

Recently a new type of electrochromic display (Schoot *et al* 1973a, b) was described utilizing an organic substance that does not have these disadvantages. The principle of this display is an oxidation–reduction reaction of diheptylviologendibromide dissolved in water. For convenience the shorthand notation $A^{2+}Br_2^-$ will be used. In a display cell there are two types of electrodes: a cathode and an anode. At the cathode the organic ion A^{2+} may take up an electron so that the radical ion $Å^+$ is formed: $A^{2+} + e \rightleftharpoons Å^+$. The dot indicates that the ion is an organic radical. This first reduction step results in a deep-blue product. If the applied voltage is too high, a second reduction step $Å^+ + e \rightleftharpoons A$ may take place giving the slightly yellowish product A. Normally this would be a serious disadvantage, but here the compound A immediately reacts with the starting material A^{2+} giving the blue radical ion again: $A + A^{2+} \rightarrow 2Å^+$. A very important point is that the blue bromide salt is insoluble in water, and therefore precipitates on the cathode. A bluish-purple layer is thus formed which adheres to the cathode.

A further requirement is a large difference in solubility of the uncoloured compound and the coloured radical ion. A high solubility of the uncoloured compound in water means a high concentration of A^{2+} ions near the cathode, so that many ions are readily available for the reduction reaction. This makes the process sufficiently fast. A low solubility of the coloured product means that a layer is formed on the cathode. The dye precipitates quickly and adheres to the cathode. This dense layer has a low electrical conductivity so that a self-healing process helps to bring about a uniform coverage of the whole cathode surface.

At the anode, of course, an oxidation–reduction reaction must also take place. Several reactions may be used for this purpose. Usually the same reaction is utilized as at the cathode, but in the reverse direction.

For a good contrast a high optical density of the coloured layer is required. The optical density depends on the thickness of the layer and on the absorption coefficient of the blue dye. The thickness of the layer is proportional to the amount of charge transported while the absorption coefficient is a material property of the dye. The absorption spectrum is shown in figure 16. Here the optical density is plotted against wavelength for a thin layer on a transparent

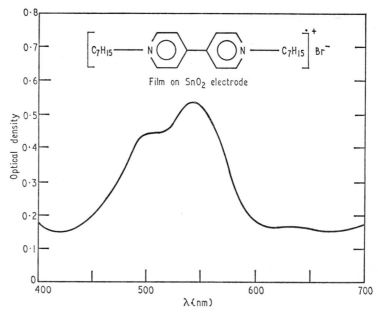

Figure 16. Absorption spectrum of diheptylviologen radical ion layer.

SnO$_2$ electrode. There is a broad absorption peak around 545 nm. The absorption coefficient is high, namely $\alpha = 26\ 000$ cm^{-1}.

A display cell utilizing the electrochromic effect, in principle, is very simple. Two glass plates with an electrode pattern are sealed together at a small distance (about 1 mm) of one another. Such a cell may be used either in a reflective mode or in a transmissive mode so that the electrodes should be reflective or transparent. Up till now Au, Pt or SnO$_2$ electrodes have been used. A specific requirement is that the electrodes should be electrochemically inert. As an oxidation–reduction reaction is used, the cell should be hermetically sealed so that no oxygen can enter. The cell is filled with an aqueous solution of the diheptylviologen compound.

The simplest type of cell uses two types of electrodes only, namely a number of working electrodes to form the character, and a common counter electrode. These electrodes can either act as cathodes or as anodes, depending on the polarity of the applied voltage. For reasons that will be discussed later, it may be advantageous to use a third type: a reference electrode. In that case the electrode configuration is as shown in figure 17. The working electrodes are in the form of 7 segments; a common counter electrode encloses the character. There are two reference electrodes, which are interconnected. In this particular cell the electrodes are in one plane.

The information is written in by applying a voltage which is larger than the redox potential, between the working electrodes and the counter electrode.

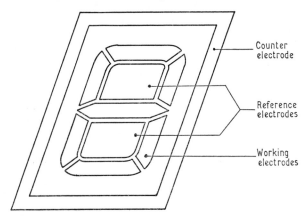

Figure 17. Electrode configuration in ECD cell.

The magnitude of this voltage determines how fast the charge required to build up the coloured layer is transferred. The higher the voltage the faster the response. As soon as sufficient contrast has been built up, that is when the coloured layer is sufficiently thick, the current source should be switched off. The coloured layer then remains on the cathode, if no oxidizing agents are present. The display therefore has memory.

The information is erased by reversing the current through the cell. This can be accomplished in three different ways. The first method is to short-circuit the working electrode and the counter electrode. In this case use is made of the potential of the layers on the electrodes and erasure is relatively slow. Application of a reverse voltage makes the erasure much faster, so that erasure times between 10 and 50 ms are obtained. A problem is, however, to know when to stop. If erasure is too long a thick coloured layer may be formed on the counter electrode, which makes the next write–erase cycle slow. The third and best method, therefore, is to use a potentiostatically controlled erasure. Now a reference electrode is used to detect the presence of a coloured layer at the working electrode by comparing the layer potential with a given potential. The erase current is adjusted to make these two potentials equal.

The writing time, of course, depends on the voltage across the cell, as is plotted in figure 18. There is a clear threshold at about 0·2 V. The value of this threshold depends on the electrode material. For these measurements gold plated nickel electrodes were used. We further note that it takes longer to make a thicker coloured layer because each radical ion requires one electron. To get an absorption of 20% takes about 2 ms; to get an absorption of 80% takes about 20 ms. To get an absorption of 80% a charge of 2 mC cm^{-2} of electrode surface has to be transferred. The speed, of course, depends on the current. A current of 100 mA cm^{-2} of electrode surface is required for a writing time of 20 ms; for a writing time of 200 ms the current is 10 mA cm^{-2}.

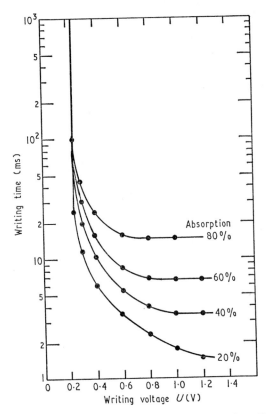

Figure 18. Writing time as a function of applied voltage U for various values of the absorption A.

An absorption of 80% means a contrast of 5 to 1. By transferring more charge a contrast of up to 20 to 1 can be obtained. It is very important that this contrast does not depend on viewing angle, just as a painted surface does not look very different if viewed from several directions.

Because the display has memory, it is only necessary to supply voltage when the information is changed. If this is infrequent then the mean power consumption is, of course, low. The peak power requirement for one write–erase cycle is about 100 mW for a character size of 1 cm² effective area and for write and erase times of about 20 ms.

With this power consumption and the low switching voltage the display is compatible with bipolar and MOST circuits. If the three-electrode configuration is used, a simple differential amplifier is needed to measure the potential difference between reference electrode and the working electrode. Multiplexing is in principle possible because the display is sufficiently fast and there is a threshold.

The life expectancy of such a display is good. A test cell has exceeded 10^7

write–erase cycles without any problem. The coloured radical ion has not shown signs of decomposition after one year of storage. There is, however, one problem left. Erasure becomes more difficult after long memory times. This means that there is a certain fatigue which still is not completely understood.

5. Gas discharge displays

A gas discharge display consists of two electrodes (or two sets of electrodes) in a sealed envelope, filled with a low pressure noble gas. A gas discharge has a nonlinear characteristic with a well defined threshold, and ignites sufficiently fast to make multiplexing and crossbar address possible. Such a display is active; the light is generated with a reasonable efficiency. The technology is flexible, so that gas discharge displays of many sizes and formats can be made. Two types may be distinguished: DC and AC displays. For a recent review see Weston (1972).

A modern DC display is a flat tube with seven-segmented numerals. The electrodes can either be made of solid metal or they can be made in a thick film technique. Such displays are usually multiple and can be driven directly from integrated circuits.

In bigger DC gas discharge displays, consisting of an array of individual discharges, the discharges are confined by aperture in an insulating plate between cathodes and anodes. In such large arrays use is made of a crossbar configuration of cathode and anode leads. The principle of the construction, as used in our laboratory, is shown in figure 19 (de Boer 1968). Panels varying in size

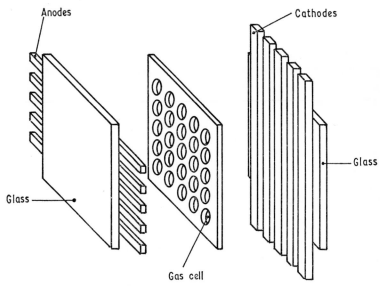

Figure 19. Principle of DC glow discharge display panel.

from 5×7 elements to 256×256 elements have been made (van Houten *et al* 1972).

To drive a big panel a considerable amount of electronic circuitry is needed and the interconnection problem may become cumbersome. An interesting idea therefore is to use SELF-SCAN†, where part of the selection is done within the panel by making use of the self-priming characteristics of neighbouring discharges. The principle of such a panel has been described by Harman (1969, 1972) and by Holtz (1972).

A different approach is used in the AC plasma panels, originating from the University of Illinois (Bitzer and Slottow 1966, 1968, Nolan 1969). Here the electrodes are isolated from the discharge by a layer of dielectric material. Such a panel is driven at a frequency of about 30 kHz. The panel has inherent memory and is therefore especially useful for the display of many characters or of graphical information. Panels up to 1024×1024 elements have been announced.

Table 3. Gas discharge displays

Applied voltage	200–300 V	DC or AC
Current/element	0·1 mA	
Switching voltage	25–75 V	
Brightness	$\leqslant 2000$ nit	Depending on size of panel
Luminous efficacy†	0·1–1 lm W^{-1}	
Response time	10 μs	Depending on priming
Threshold	Yes	

† The value of 1 lm W^{-1} is for a colour gas discharge display making use of miniature positive column discharges.

All these gas discharge displays use glow discharges and therefore have the well known orange-red neon colour. Recent work has shown that in principle multicolour displays can be made (see van Gelder and Matthey 1973, Forman 1972, Brown and Tamm Zayac 1972).

All gas discharge displays have relatively high ignition voltages, say of 200–300 V, and maintaining voltages of 150–250 V. The currents are relatively low (see table 3). It is often possible to apply a bias voltage so that the switching voltages can be 25 to 75 V. These displays therefore can be directly driven from integrated circuits. In very big display panels the current per line is, nevertheless, often so high that the line drivers still have to be made with discrete transistors. The brightness of a single gas discharge cell can be very high (see table 3). In big panels the mean brightness is, of course, lower because light emission occurs for only a short period during each address cycle. A mean brightness of 150 nit for a panel of 400 cm^2 is feasible.

† SELF-SCAN is a registered trade name of Burroughs Corporation.

6. Light-emitting diodes

The last type of display to be discussed makes use of light-emitting diodes (LEDS). Some recent reviews have been written by Thomas (1971), by Bergh and Dean (1972) and by Loebner (1973). A LED is a forward-biased p–n junction diode. Light is generated owing to recombination of injected charge carriers near the p–n junction. The overall efficacy of this process is determined by four factors:

(i) the injection efficiency, which is in the order of 25 to 50%;

(ii) the conversion efficiency, which strongly depends on the material used;

(iii) the optical losses, due to absorption and multiple internal reflections. The latter effect can be reduced by shaping the diode and by plastic encapsulation;

(iv) the eye sensitivity for the emitted radiation, which depends strongly on the wavelength of the emitted light. The human eye is most sensitive in the green and much less in the red and in the blue.

To enable visible light to be emitted, the band gap of the semiconductor material used to make the diodes should be between 1·7 and 3·3 eV. Potentially suitable materials may be found among the II–VI and the III–V compounds. It has hitherto been impossible to make p–n junctions in II–VI compounds, although the use of ion implantation techniques and of MIS structures is opening up new possibilities. Best results have been achieved using the III–V materials GaP and GaAsP, although some other III–V materials are at present being investigated.

The GaAsP diodes emit in the red. They can be made relatively easily by gas-phase epitaxy and a diffusion process. They are cheap. Their power efficiency, however, is only about 0·1% (at 650 nm). Yellow GaAsP:N lamps have recently become available (see table 4).

Red GaP diodes are made by liquid-phase epitaxy, which is a more expensive process. Green diodes are made by liquid or by vapour-phase epitaxy or by a

Table 4. Light-emitting diodes. (Data are for commercially available devices. Research values are higher)

Material	GaP	GaAsP
Technology	LPE VPE (green)	VPE
Colour	Red 690 nm Green 570 nm (Yellow)	Red 650 nm Yellow 590 nm
Power efficiency	Red 0·5–2% Green 0·05%	Red 0·1% Yellow 0·04%
Eye sensitivity	Red 20 lmW^{-1} Green 610 lmW^{-1}	Red 75 lmW^{-1} Yellow 450 lmW^{-1}
Luminous efficacy	Red 0·1–0·4 lmW^{-1} Green 0·06 lmW^{-1}	Red 0·07 lmW^{-1} Yellow 0·04 lmW^{-1}

combination of both processes. The commercially available diodes have power efficiencies of up to 2% for the red lamps, and about 0·05% for the green diodes. If the green diodes are pulsed with high currents the efficiency can be much higher. In several research laboratories higher efficiencies have been realized. The highest reported value for the red GaP diodes is 15%. Yellow diodes have also been made (see table 4).

GaAs diodes are very efficient infrared emitters and require an anti-Stokes phosphor for visible emission. In this way several colours can be made. The overall efficacy is low at moderate currents, especially for the blue where a three-step phosphor is required. Now that green and yellow LEDs can be made, the interest in this diode–phosphor combination seems to diminish.

LEDs are low impedance devices: they work at low voltages, but at high currents. They will, therefore, not be used in large arrays, but in α-numeric indicators. The luminous efficacies, expressed in $lm\,W^{-1}$, are still lower than that of a glow discharge display. It is to be expected that commercially available devices will reach the higher power efficiencies that are now achieved in the laboratory. There is still the problem that the best diodes emit in the red, where the human eye is not very sensitive. It is better with the green and yellow lamps, but there the power efficiency is low.

The material and the processing is still rather expensive. Therefore most displays are hybrid devices, consisting of individual diodes mounted on a substrate, usually arranged in a 7-segmented format. The dimensions of such a display are usually small, to keep the materials cost down. Simple lenses are often used to give larger characters. Recently monolithic displays have been made, which are more expensive because more material is needed, but have better performance and longer life.

LEDs can be multiplexed, because they are fast and have a threshold. The red GaP diodes, however, saturate at higher currents so that strobing is restricted to a few digits. This problem is not present with the green GaP lamps or with GaAsP diodes, where the current is limited by burn-out only.

7. Comparison of displays

Let us now see how the four display principles compare. In table 5 some important display parameters have been summarized. Additionally, Gordon and Anderson (1973) have recently discussed a number of display technologies.

Gas discharge and LED give an active display; liquid crystals and electro-chromic displays are passive.

Glow discharge displays work at high voltages, but at low currents. The luminous efficacy is reasonable. The brightness can be high. They can be multiplexed and crossbar address of large matrix displays is possible.

Light-emitting diodes are low impedance devices. They work at low voltage, but the currents are high. The brightness can be high and it is to be

Table 5. Comparison of five types of displays

	Glow discharge	Light emitting diodes	'Dynamic scattering' in liquid crystals	Field effects in liquid crystals	Electro-chromic
Display	Active	Active	Passive	Passive	Passive
Voltage	200–300 V	2–5 V	20 V	1–100 V	1 V
Current/ character	0·1–1·0 mA	10–100 mA	10 μA	1 μA	10–100 mA†
Luminance	⩽2000 nit	⩽2000 nit	Passive	Passive	Passive
Luminous efficacy	0·1–1·0 lm W^{-1}	0·1–0·4 lm W^{-1}	Not applicable	Not applicable	Not applicable
Response time	10 μs	100 ns	10–100 ns	30 μs– 100 ms	20–200 ms†
Threshold	Yes	Yes	No	Yes	Yes
Multiplexing	Yes	Yes, but. . .	No, but. . .	Yes	Yes, but. . .
Large matrix displays	Yes	No	No	No	No
Memory	Possible	No	No	Possible	Yes

† In an electrochromic display the current depends on the response time that is required and flows only during writing and erasing.

expected that the luminous efficacies will further improve in the years to come. They can be multiplexed, but the number of digits to be strobed is limited. They are not suitable for large matrix displays.

Liquid crystals consume very little power. Multiplexing of dynamic scattering displays is difficult. The field effects are more promising in this respect. Large matrix displays, however, cannot be made with liquid crystals. The life problems have been solved in the laboratory, but there is little experience as yet with commercial samples.

Electrochromic displays also consume little power, they have a good contrast and have memory. They are not suitable for matrix displays. The work is still at the laboratory stage.

A question that is often posed is, which display technology looks the most promising? In our opinion, there is not one type of display that can fulfil the requirements of all applications. In table 1 the whole display field is divided according to the size of the display panel. From left to right the following general tendencies may be noted:

(i) the interconnection problem is becoming increasingly important;
(ii) the same applies to the need to integrate the driving circuitry;
(iii) a free choice of colour or even multi-colour is needed.

At the right-hand side of the table the cathode ray tube can fulfil the requirements rather well. It is likely to remain the best solution for many applications where much information has to be displayed. When the number of characters decreases, the cathode ray tube gradually becomes an expensive and cumbersome solution.   That is where the gas discharge panels may take over.   They have the advantage of digital address, flat display and high brightness.   The technology of these gas discharge panels is flexible, so that displays of many sizes and formats can be made.   The applicability of gas discharge from intermediate size panel displays extends to single character tubes.

To fulfil the requirements at the left-hand side of the table, many competing technologies have been proposed.   Many applications are in the consumer market, and price, therefore, is a very important factor.   Liquid crystals are potentially cheap and consume little power.   This makes them especially suitable for battery-operated, portable apparatus.   The price per character hardly depends on character size, so that they may be used for big characters as well.   Electrochromic displays are a newcomer in the field and although very promising it is still too early to assess their potentiality with any accuracy. They are especially suitable for those applications where the information is infrequently changed.   LEDs are very reliable, but they are still relatively expensive and price goes linearly up with size.   They will therefore mainly be used as numeric indicators for the display of small characters.

Acknowledgments

The author wishes to thank his colleagues at the Philips Research Laboratories for their help and criticism, especially Dr C Z van Doorn for his permission to make use of an internal report on liquid crystals.

References

Alburger J R 1957 *Electron. Inds.* Feb p 50
Assouline G, Hareng M, Leiba F and Roneillat M 1971 *Electron. Lett* **7** 699
Bergh A A and Dean P J 1972 *Proc. IEEE* **60** 156
Bitzer D L and Slottow H J 1966 *paper presented at Fall Joint Computer Conf.* p 541
—— 1968 *paper presented at IEEE Microelectr. Symp.*
de Boer Th J 1968 *paper presented at 9th Nat. Symp. on Inf. Display, Los Angeles* p 193
Brown F H and Tamm Zayac M 1972 *Proc. SID* **13** 52
Chistiakov I G 1967 *Sov. Phys.–Usp.* **9** 551
Deb S K 1969 *Appl. Opt. Suppl.* **3** 192
van Doorn C Z 1973 *Phys. Lett.* **42A** 537
Fergason J L 1964 *Scient. Amer.* Aug p 77
Forman J 1972 *Proc. SID* **13** 14
van Gelder Z and Matthey M M M P 1973 *Proc. IEEE* **61** 1019
Gerritsma C J, De Jeu W H and van Zanten P 1971 *Phys. Lett.* **36A** 389
Gerritsma C J and Lorteye J H J 1973 *Proc. IEEE* **61** 829
Gooch C H and Low J J 1972 *Appl. Phys.* **5** 1218
Gordon E I and Anderson L K 1973 *Proc. IEEE* **61** 807
Harman W J jr 1969 *paper read at IEEE Int. Electron. Dev. Mtg., Washington*
—— 1972 *IEEE Int. Conv. Techn. Dig.* 72

Heilmeyer G H and Goldmacher J E 1968 *Appl. Phys. Lett.* **13** 132
—— 1969 *Proc. IEEE* **57** 134
Heilmeyer G H, Zanoni L A and Barton L A 1968 *Proc. IEEE* **56** 1162
Holtz G E 1972 *Proc. SID* **13** 2
van Houten S, Jackson R N and Weston G F 1972 *Proc. SID* **13** 43
van Houten S and de Vries G H F 1972 *Proc. 1st Europ. Electro-Optics Markets and Techn. Conf., Geneva* p 324
Jakeman E and Raynes E P 1972 *Phys. Lett.* **39A** 69
de Jeu W H, Gerritsma C J and Lathouwers T W 1972 *Chem. Phys. Lett.* **14** 503
de Jeu W H and van der Veen J 1972 *Philips Res. Rep.* **27** 172
—— 1973 *Phys. Lett.* **44A** 277
Kahn F J 1972 *Appl. Phys. Lett.* **20** 199
Kahn F J, Taylor G N and Schonhorn H 1973 *Proc. IEEE* **61** 823
Leslie F M 1970 *Molec. Cryst. Liq. Cryst.* **12** 57
Loebner E E 1973 *Proc. IEEE* **61** 837
Luxemberg H R and Kuhn R L (ed) 1968 *Display Systems Engineering* (New York: McGraw-Hill)
Nolan J F 1969 *paper read at Int. Electr. Dev. Conf., Washington*
Orsay Liquid Crystal Group 1971 *Molec. Cryst. Liq. Cryst.* **12** 251
Schadt M and Helfrich W 1971 *Appl. Phys. Lett.* **18** 127
Schiekel M F and Fahrenschon K 1971 *Appl. Phys. Lett.* **19** 391
Schoot C J, Bolwijn P T, van Dam H T, van Doorn R A, Ponjeé J J and van Houten S 1973a *paper presented at SID Int. Symp. NY*
Schoot C J, Ponjeé J J, van Dam H T, van Doorn R A and Bolwijn P T 1973b, *J. Appl. Phys.* **23** 64
Soref R A and Rafuse M J 1972 *J. Appl. Phys.* **43** 2029
Stein C R and Kashnow R A 1971 *Appl. Phys. Lett.* **19** 343
Sussman A 1972 *IEEE Trans. Parts Hybrids Pack* **PHP8** 24
Thomas D G 1971 *IEEE Trans.* **ED-18** 621
Weston G F 1972 *Glow Discharge Display* (London: Mills and Boon)
Wild P J and Nehring J 1971 *Appl. Phys. Lett.* **19** 335
Wysocki J, Adams J and Haas W 1969 *Molec. Cryst. Liq. Cryst.* **8** 471
Zaromb S 1962 *J. Electrochem Soc.* **109** 903